ADMISSIONS

Also by Henry Marsh

Do No Harm

ADMISSIONS

Life as a Brain Surgeon

Henry Marsh

Thomas Dunne Books
St. Martin's Press
New York

THOMAS DUNNE BOOKS.
An imprint of St. Martin's Press.

www.thomasdunnebooks.com
www.stmartins.com

The Library of Congress Cataloging-in-Publication Data
is available upon request.

ISBN 978-1-250-12726-6 (hardcover)
ISBN 978-1-250-12727-3 (ebook)

Our books may be purchased in bulk for promotional, educational, or business use. Please contact your local bookseller or the Macmillan Corporate and Premium Sales Department at 1-800-221-7945, extension 5442, or by email at MacmillanSpecialMarkets@macmillan.com.

First published in Great Britain by Weidenfeld & Nicolson, an imprint of The Orion Publishing Group Ltd, an Hachette UK company

First U.S. Edition: October 2017

10 9 8 7 6 5 4 3 2 1

To William, Sarah, Katharine and Iris

'Neither the sun nor death can be looked at steadily'

La Rochefoucauld

'We should always, as near as we can, be booted and spurred, and ready to go . . .'

Michel de Montaigne

'Medicine is a science of uncertainty, and an art of probability . . .'

Sir William Osler

CONTENTS

PREFACE

I like to joke that my most precious possession, which I prize above all my tools and books, and the pictures and antiques that I inherited from my family, is my suicide kit, which I keep hidden at home. It consists of a few drugs that I have managed to acquire over the years. But I don't know whether the drugs would still work – they came with neither a 'Use By' nor a 'Best Before' date. It would be embarrassing to wake up in Intensive Care after a failed suicide attempt, or to find myself having my stomach pumped out in Accident and Emergency. Attempted suicides are often viewed by hospital staff with scorn and condescension – as failures in both living and dying, and as the agents of their own misfortune.

There was a young woman, when I was a junior doctor and before I started training to be a brain surgeon, who was saved from a barbiturate overdose. She had been determined to die in the wake of an unhappy love affair, but had been found unconscious by a friend and taken to hospital, where she was admitted to the ITU – the Intensive Therapy Unit – and ventilated for twenty-four hours. She was then transferred to the ward where I was a houseman – the most junior grade of hospital doctor – when she started to wake up. I watched her regain consciousness, coming back to life, surprised and puzzled at first still to be alive, and then not quite sure whether she wanted to return to the land of the living or not. I remember sitting on the edge of her bed and talking with her. She was very thin, and was obviously anorexic. She had short, dark-red hair, which was matted and dishevelled after a day in a coma on a ventilator. She sat with her chin resting on the hospital blanket over her drawn-up knees. She was quite calm; perhaps this was still the effect of the overdose, or perhaps it was because she felt that here, in hospital, she was in limbo, between heaven and hell – that she had been given a brief reprieve from her unhappiness. We became friends of a kind for the two days that she was on the ward and before she was transferred to the care of the psychiatrists. It turned out that we had acquaintances in common from Oxford in the past, but I do not know what happened to her.

I have to admit that I'm not at all sure that I would ever dare to use the drugs in my suicide kit when – and it may well happen quite soon – I am faced with the early signs of dementia, or if I develop some incurable illness such as one of the malignant brain tumours with which I am so familiar from my work as a brain surgeon. When you are feeling fit and well, it is relatively easy to entertain the fantasy of dying with dignity by taking your own life, as death is still remote. If I don't die suddenly, from a stroke or a heart attack, or from being knocked off my bicycle, I cannot predict what I will feel when I know that my life is coming to an end – an end which might well be distressing and degrading. As a doctor, I cannot have any illusions. But it wouldn't entirely surprise me if I started to cling desperately to what little life I had left. Apparently, in countries where so-called doctor-assisted suicide is legal many people, if they have a terminal illness, having initially expressed an interest in being able to die quickly, do not take up the option as the end approaches. Perhaps all that they wanted was the reassurance that if the end was to become particularly unpleasant, it could be brought to a quick conclusion and, in the event, their final days passed peacefully. But perhaps it was because, as death approached, they started to hope that they might yet still have a future. We develop what psychologists call 'cognitive dissonance', where we entertain entirely contradictory thoughts. Part of us knows, and accepts, that we are dying but another part of us feels and thinks that we still have a future. It is as though our brains are hardwired for hope, or at least that part of them is.

As death approaches, our sense of self can start to disintegrate. Some psychologists and philosophers maintain that this sense of self, of being coherent individuals free to make choices, is little more than a title page to the great musical score of our subconscious, a score with many obscure, often dissonant voices. Much of what we think of as real is a form of illusion, a consoling fairy story created by our brains to make sense of the myriad stimuli from inside and outside us, and of the unconscious mechanics and impulses of our brains.

Some even claim that consciousness itself is an illusion – that it is not 'real', that it is a trick played on us by our brains – but I do not understand what they mean by this. A good doctor will speak to both the dissonant selves of a dying patient – the part that knows that it is dying, and the part that hopes that it will yet live. A good doctor will neither lie nor deprive the patient of hope, even if the hope is only of life for a few more days. But it is not easy, and it takes time, with many long silences. Busy hospital wards – where most of us are still doomed to die – are not good places in which to have such conversations. As we lie dying, many of us will keep a little fragment of hope alive in a corner of our minds, and only near the very end do we finally turn our face to the wall and give up the ghost.

I

THE LOCK-KEEPER'S COTTAGE

The cottage stands on its own by the canal, derelict and empty, the window frames rotten and hanging off their hinges and the garden a wilderness. The weeds were as high as my chest and hid, I was to discover, fifty years of accumulated rubbish. It faces the canal and the lock, and behind it is a lake, and beyond that a railway line. The property company that owned it must have paid somebody to clear out the inside of the cottage, and whoever had done the work had simply thrown everything over the old fence between the garden and the lake, so the lake side was littered with rubbish – a mattress, a disembowelled vacuum cleaner, a cooker, legless chairs and rusty tins and broken bottles. Beyond the junk, however, lay the lake, lined by reeds, with two white swans in the distance.

I first saw the cottage on a Saturday morning. A friend had told me about it. She had seen that it was for sale and knew that I was looking for a place where I could establish a woodworking workshop in Oxford to help me cope with retirement. I parked my car beside the bypass and walked along the flyover, deafening cars and trucks rushing past me, to find a small opening in the hedge, almost invisible, at the side of the road. There was a long line of steps covered in leaves and beechmast, under a dark archway formed by the low, bending branches of beech trees, leading down to the canal. It was as though I was suddenly dropping out of the present and returning to the past. The roar of the traffic became abruptly muted as I descended to the quiet and still canal. The cottage was a few hundred yards away along the towpath, over an old, brick-built humpback canal bridge.

There were several plum trees in the garden, one of them growing up through an obsolete and rusty old machine with reciprocating blades like a hedge-trimmer, for cutting heavy undergrowth. It had two big wheels with *Allens* and *Oxford* stamped on the rims in large letters. My father had had exactly the same model of machine, which he used in the two-acre garden and orchard where I had grown up less than one mile away in the 1950s. He once accidentally ran over a little shrew in the grass of the orchard as I stood watching him, and I remember my distress at seeing its bleeding body and hearing its piercing screams as it died.

The cottage looks out over the still and silent canal and the heavy black gates of the narrow lock. There is no road access – it can only be reached along the towpath on foot or by barge. There is a brick wall with drinking troughs for horses along one side of the garden, facing the canal – I found later the metal rings to which the horses which towed the barges along the canal would have been tethered. A long time ago the lock-keeper would have been responsible for the gates, but the lock-keepers' cottages along the canal have all been sold off and the gates are now left to be operated by whoever is on the passing barges. I am told that a kingfisher lives here and can be seen flashing across the water, and that there are otters as well, even though only a few hundred yards away there is the roar of the bypass traffic crossing the canal on the high flyover on its concrete stilts. But if I turn away from the road, all I can see are fields and trees, and the reed-lined lake behind the house. I can imagine that I am in ancient, deep countryside, as it was when I was growing up nearby, before the bypass was built sixty years ago.

The young woman from the estate agents was sitting on the grass bank in the sunshine beside the entrance to the cottage, waiting for me. She opened the bolted and padlocked front door. I stepped over a few letters on the floor inside, covered in muddy footprints. The estate agent saw me looking down at them and told me that an old man had lived here by himself for almost fifty years – the deeds for the property described him as a canal labourer. When he died the property developers, who had bought the house some years ago, put it up for sale. She did not know whether he had died here or in hospital or in a nursing home.

The place smelt damp and neglected. The cracked and broken windows were covered by torn, dirty lace curtains and the window sills were black with dead flies. The rooms had been stripped out and had the sad and despondent air of all abandoned homes. Although there was water and electricity, the facilities were primitive, and there was only an outside toilet, smashed into pieces, with the door off its hinges. The dustbin by the front door contained plastic bags full of faeces.

The ancient farmouse nearby where I had spent my childhood was said to have been haunted – at least, according to the Whites, the elderly couple who lived across the road and whom I liked to visit. An improbable tale of a sinister coach and horses in the yard at night and also of a 'grey lady' in the house itself. It was easy to imagine the old man's ghost haunting the cottage.

'I'll take it,' I said.

The girl from the estate agents looked at me sceptically.

'But don't you want to get a survey?'

'No, I do all my own building work and it looks OK to me,' I replied confidently, but wondering whether I was still capable of the physical work that would be required and how I would manage without any road access. Perhaps I should stop being so ambitious and abandon my obsessive conviction that I must do everything myself. Perhaps it no longer mattered. I ought to employ a builder. Besides, although I wanted a workshop, I wasn't sure that I wanted to live in this small and lonely cottage, with a possible ghost.

'Well, you'd better make an offer to Peter, the manager in our local office,' she replied.

I drove back to London the next day – with the uneasy thought that perhaps this little cottage would be where I myself would eventually end my days and die, and where my story would end. Now that I am retiring, I am starting all over again, I thought, but now I am running out of time.

I was back in the operating theatre on Monday – I was in my blue theatre scrubs, but expected to be only an observer. In three weeks' time I was to retire – after almost forty years of medicine and neurosurgery. My successor, Tim, who had started off as a trainee in our department, had already been appointed. He is an exceptionally able and nice man, but not without that slightly fanatical determination and attention to detail that neurosurgery requires. I was more than happy to be replaced by him and it seemed appropriate to leave most of the operating to him, in preparation for the time when – and it would probably be something of a shock for him – he suddenly carried sole responsibility for what happened to the patients under his care.

The first case was an eighteen-year-old woman who had

been admitted for surgery the previous evening. She was five months pregnant but had started to suffer from severe headaches, and a scan showed a very large tumour – almost certainly benign – at the base of her brain. I had seen her as an emergency in my outpatient clinic a few days earlier; she came from Romania and her English was limited, but she smiled bravely as I tried to explain things to her via her husband, who spoke a little English. He told me that they came from Maramures, the area of northern Romania on the border with Ukraine. I had been there myself two years ago on a journey from Kiev to Bucharest with my Ukrainian colleague Igor. The landscape was exceptionally beautiful, with ancient wooden farms and monasteries – it seemed that the modern world had scarcely caught up with the place at all. There were haystacks in the fields and hay wagons drawn by horses on the roads, with the drivers wearing traditional peasant costumes. Igor was outraged that Romania had been allowed to join the European Union whereas Ukraine had been kept out. My Romanian colleague, who had come to collect us from the border with Ukraine, wore a tweed cloth cap and leather driving gloves, and drove us at high speed on the terrible roads in his son's souped-up BMW all the way to Bucharest, almost without stopping. We did, however, spend a night on the way at Sighisoara, where the house still stood where Vlad the Impaler – the prototype for Dracula – had been born. It was now a fast-food joint.

The operation on the woman was not an emergency in the sense that it did not need to be done at once, but it certainly had to be done within a matter of days. Such cases do not fit easily into the culture of targets which now defines how the National Health Service in England is supposed to function. She was not a routine case but nor was she an emergency.

My own wife Kate, a few years ago, had fallen into the

same trap when awaiting major surgery after many weeks of intensive care at a famous hospital. She had been admitted as an emergency and underwent emergency surgery without any difficulty, but then needed further surgery after several weeks of intravenous feeding. I became accustomed to the sight of a large foil-wrapped bag of glutinous fluid hanging above her bed, dripping into her central line – a catheter inserted into the great veins leading to her heart. Kate was now no longer an emergency but nor was she a routine admission, so there was no provision for her to undergo surgery. For five days in a row she was prepared for surgery – very major surgery, with all manner of frightening potential complications – and each day by midday the operation was cancelled. Eventually, in despair, I rang her surgeon's secretary. 'Well, it's not really up to Prof as to who goes on the routine operating lists,' she explained apologetically. 'It's a manager – the *List Broker*. Here's the number to ring . . .'

So I rang the number only to receive a message that the voice mailbox was full and I could not leave a message. At the end of the week the decision was made to make Kate into a routine case by sending her home with a large bottle of morphine. She was readmitted a week later, presumably now with the List Broker's permission. The operation was a great success, but I mentioned the problem we had encountered to one of my neurosurgical colleagues at the same hospital when we met at a meeting shortly afterwards.

'I find it very difficult being a medical relative,' I said. 'I don't want people to think my wife should get better treatment just because I'm a surgeon myself, but it really was getting pretty unbearable. Having your operation cancelled is bad enough – but five days in a row!'

My colleague nodded. 'And if we can't look after our own, what about Joe Bloggs?'

So I had gone to work on Monday morning worried that there would be the usual shambles of trying to find a bed for the young girl into which she could go after surgery. If her condition was life-threatening I would be able to start the operation without having to seek the permission of the many hospital staff involved in trying to allocate an insufficient number of beds to too many patients, but her condition was not life-threatening – at least not yet – and I knew that I was going to have a difficult start to the day.

At the theatre reception area there was an animated group of doctors and nurses and managers looking at the day's operating lists sellotaped to the top of the desk, discussing the impossibility of getting all the work done. I saw that several of the cases were routine spinal operations.

'There are no ITU beds,' the anaesthetist said with a grimace.

'Well why not just send for the patient anyway?' I asked. 'A bed always turns up later.' I always say this, and always get the same reply.

'No,' she said. 'If there's no ITU bed I will end up having to recover the patient in theatre after the op and it could take hours.'

'I'll try to go and sort it out after the morning meeting,' I replied.

There was the usual collection of disasters and tragedies at the morning meeting.

'We admitted this eighty-two-year-old man with known prostate cancer yesterday. He had gone first to his local hospital because he was going off his legs and was in retention of urine. They wouldn't admit him and sent him home,' Fay, the on-call registrar, told us as she put up a scan. This was met with sardonic laughter in the darkened room.

'No, no, it's true,' Fay said. 'They catheterized him and

wrote in the notes that he was now much better. I have seen the notes.'

'But he couldn't fucking walk!' somebody shouted.

'Well, that didn't seem to trouble them. At least they must have achieved their four-hour target by sending him home. He spent forty-eight hours at home and the family got the GP in, who sent him here.'

'Must have been a very uncomplaining and long-suffering patient,' I observed to my colleague sitting next to me.

'Samih,' I said to one of the other registrars, 'what do you see on the scan?' I had first met Samih some years earlier on one of my medical visits to Khartoum. I had been very impressed by him and did what I could to help him to come to England to continue his training. In the past it had been relatively easy to bring trainees over to my department from other countries, but the combination of European Union restrictions on doctors from outside Europe and increasing bureaucratic regulations in recent years has made it very difficult, even though the UK has fewer doctors per capita than any country in Europe other than Poland and Romania. Samih passed all the required examinations and hurdles with flying colours. He was a joy to work with, a large and very gentle man, utterly dedicated to our craft, who was loved by the patients and nurses. He was now to be my last registrar.

'The scan shows metastatic posterior compression of the cord at T3. The rest of the scan looks OK.'

'What's to be done?' I asked.

'Well, it depends on how he is.'

'Fay?'

'He was sawn off when I saw him at ten o'clock last night.'

This is the brutal but accurate phrase to describe a patient who has a spinal cord so badly damaged that they have no feeling or movement of any kind below the level of the damage and when there is no possibility of recovery. T3 means the third thoracic vertebra, so the poor old man would have no movement of his legs or trunk muscles. He would even have difficulties just trying to sit upright.

'If he's sawn off he's unlikely to get better,' Samih said. 'It's too late to operate now. It would have been a simple operation,' he added.

'What's this man's future?' I asked the room at large. Nobody replied so I answered the question myself.

'It's very unlikely he'll be able to get home as he'll need full twenty-four-hour nursing, with being turned every few hours to prevent bed sores. It takes several nurses to turn a patient, doesn't it? So he will be stuck in some geriatric ward somewhere until he dies. If he's lucky the cancer elsewhere in his body will carry him off soon, and he may make it into a hospice first, nicer than a geriatric ward, but the hospices won't take people if their prognosis is that they might live for more than a few weeks. If he's unlucky, he may hang on for months.'

I wondered if that was how the old man in the cottage had died, alone in some impersonal hospital ward. Would he have missed his home, the little cottage by the canal, even though it was in such a sorry state? My trainees are all much younger than I am; they still have the health and self-confidence of youth, which I too had at their age. As a junior doctor you are pretty detached from the reality that faces so many of the older patients. But now I am losing my detachment from patients as I prepare to retire. I will become a member of the underclass of patients – as I was before I became a doctor, no longer one of the elect.

The room remained silent for a while.

'So what happened?' I asked Fay.

'He came in at ten in the evening and Mr C. planned to

operate but the anaesthetists refused – they said there was no prospect of his getting better and they weren't willing to do it at night.'

'Well, there's not much to be lost by operating – we can't make him any worse,' somebody said from the back of the room.

'But is there any realistic prospect of making him better?' I asked, but I went on to say: 'Although, to be honest, if it was me I'd probably say go and operate . . . just in case . . . The thought of ending my days paraplegic on a geriatric ward is so awful . . . indeed, if the operation killed me, I wouldn't complain.'

'We decided to do nothing,' Fay said. 'We're sending him back to his local hospital today – if there's a bed there, that is.'

'Well, I hope they take him back – we don't want another Rosie Dent.' Rosie had been an eighty-year-old woman earlier in the year with a cerebral haemorrhage whom I had been forced to admit by a physician at my own hospital – at least, so many complaints and threats were made if I didn't admit her to an acute neurosurgical bed that I gave in – even though she did not need neurosurgical treatment. It proved impossible to get her home and she sat on the ward for seven months, before we eventually managed to persuade a nursing home to accept her. She was a charming, uncomplaining old lady and we all became quite fond of her, even though she was 'blocking' one of our precious acute neurosurgical beds.

'I think it will be OK,' Fay said. 'It's only our own hospital which refuses to take patients back from the neurosurgical wards.'

'Any other admissions?' I asked.

'There's Mr Williams,' Tim said. 'I was hoping to do him at the end of your list after the girl with a meningioma.'

'What's the story?' I asked.

'He's had some epileptic fits. Been behaving a bit oddly of late. Used to be pretty high-functioning – engineer or something like that. Fay, could you put the scan up please?'

The scan flashed up on the wall in front of us. 'What's it show, Tiernan?' I asked one of the most junior doctors, known as SHOs, short for senior house officer.

'Something in the left frontal lobe.'

'Can you be a bit more precise? Fay, put up the Flair sequence.'

Fay showed us some different scan images, sequences that are good for indicating tumours which are invading the brain rather than just displacing it.

'It looks as though it's infiltrating all of the left frontal lobe and most of the left hemisphere,' Tiernan said.

'Yes,' I replied. 'We can't remove the tumour, it's too extensive. Tiernan, what are the functions of the frontal lobes?'

Tiernan hesitated, finding it hard to reply.

'Well, what happens if the frontal lobes are damaged?' I asked.

'You get personality change,' he replied immediately.

'What does that mean?'

'They become disinhibited – get a bit knocked off . . . ', but he found it difficult to describe the effects in any more detail.

'Well,' I said, 'the example of disinhibition loved by doctors is the man who pisses in the middle of the golfing green. But the frontal lobes are where all our social and moral behaviour is organized. You get a whole variety of altered social behaviours if the frontal lobes are damaged – almost invariably for the worse. Sudden outbursts of violence and irrational behaviour are among the commonest. People who were previously kind and considerate become coarse and selfish, even though their intellect can be perfectly well preserved. The person with frontal-lobe damage rarely has any insight into it – how can the "I" know that it is changed? It has nothing to compare itself with. How can I know if I am the same person today as I was yesterday? I can only assume that I am. Our selves are unique and can only know ourselves as we are now, in the immediate present. But it's terrible for the families. They are the real victims. Tim, what do you hope to achieve?'

'If we take some of it out, create some space, we'll buy him a bit more time,' Tim replied.

'But will surgery get his personality change any better?'

'Well, it might,' Tim said. I was silent for a while.

'I rather doubt it,' I eventually commented. 'But it's your case. And I haven't seen him. Did you discuss all this with him and his family?'

'Yes.'

'It's nine o'clock,' I said. 'Let's see what's happening about beds and find out if we are allowed to start operating.'

An hour later, Tim and Samih started the operation on the Romanian woman. I spent most of the time sitting on a stool, my back propped up against the wall behind me, while Tim and Samih slowly removed the tumour. The lights in the theatre were dimmed as they were using the microscope, and I dozed, listening to the familiar sounds and muted drama of the theatre – the bleeping of the anaesthetic monitors, the sighing of the ventilator, Tim's instructions to Samih and the scrub nurse Agnes and the hiss of the sucker which Tim was using to suck the tumour out of the woman's head. 'Toothed forceps . . . Adson's . . . diathermy . . . Agnes, pattie please . . . Samih, can you suck here? . . . there's a bit of a bleeder . . . ah! got it . . .'

I could also hear the quiet conversation between the two

anaesthetists at the far end of the table, where they sat on stools next to the anaesthetic machine with its computer screen showing the girl's vital functions, as they are called – the functioning of her heart and lungs. These appear as a series of pretty, bright-coloured lines and numerals in red and green and yellow. In the distance, from the prep area between the theatres, there would be occasional bursts of laughter and chatter from the nurses – all good friends of mine, with whom I had been working for many years – as they prepared the instruments for the next cases.

Will I miss this? I asked myself. This strange, unnatural place that has been my home for so many years, a place dedicated to cutting into living bodies and, in my case, the human brain – windowless, painfully clean, air-conditioned and brilliantly lit, with the operating table in the centre, beneath the two great discs of the operating lights, surrounded by machines? Or when the time comes in a few weeks, will I just walk away without any regrets at all?

A long time ago, I thought brain surgery was exquisite – that it represented the highest possible way of using both hand and brain, of combining art and science. I thought that brain surgeons – because they handle the brain, the miraculous basis of everything we think and feel – must be tremendously wise and understand the meaning of life. When I was younger I had simply accepted the fact that the physical matter of brains produces conscious thought and feeling. I thought the brain was something that could be explained and understood. As I have got older, I have instead come to realize that we have no idea whatsoever as to how physical matter gives rise to consciousness, thought and feeling. This simple fact has filled me with an increasing sense of wonder, but I have also become troubled by the knowledge that my brain is an ageing organ, just like the organs of the rest of my body. That my 'I' is ageing and that I have no way of knowing how it might have changed. I look at the liver spots on the wrinkled skin of my hands, the hands whose use has been the dominant theme of my life, and wonder what my brain would look like on a brain scan. I worry about developing the dementia from which my father died. On the brain scan that was done some years before his eventual death, his brain had looked like a Swiss cheese – with huge holes and empty spaces. I know that my excellent memory is no longer what it was. I often struggle to remember names.

My understanding of neuroscience means that I am deprived of the consolation of belief in any kind of life after death and of the restoration of what I have lost as my brain shrinks with age. I know that some neurosurgeons believe in a soul and afterlife, but this seems to me to be the same cognitive dissonance as the hope the dying have that they will yet live. Nevertheless, I have come to find a certain solace in the thought that my own nature, my I – this fragile, conscious self writing these words that seems to sail so uncertainly on the surface of an unfathomable, electrochemical sea into which it sinks every night when I sleep, the product of countless millions of years of evolution – is as great a mystery as the universe itself.

I have learnt that handling the brain tells you nothing about life – other than to be dismayed by its fragility. I will finish my career not exactly disillusioned but, in a way, disappointed. I have learnt much more about my own fallibility and the crudity of surgery (even though it is so often necessary), than about how the brain really works. But as I sat there, the back of my head resting against the cold, clean wall of the operating theatre, I wondered if these were just the tired thoughts of an old surgeon about to retire.

The woman's tumour was growing off the meninges – the thin, leathery membrane that encases the brain and spinal cord – in the lower part of the skull known as the posterior cranial fossa. It was immediately next to one of the major venous sinuses. These are drainpipe-like structures that continuously drain huge volumes of deep-purple, deoxygenated blood – blood which would have been brilliant red when it first reached the brain, pumped up from the heart. Blood flashes through the brain in a matter of seconds, one quarter of all the blood from the heart, darkening as the brain takes the oxygen out of it. Thinking, perceiving and feeling, and the control of our bodies, most of it unconscious, are energy-intensive processes fuelled by oxygen. There was some risk that removing the tumour might tear the transverse venous sinus and cause catastrophic haemorrhage, so I scrubbed up and helped Tim with the last twenty minutes of the operation, carefully burning and peeling the tumour off the side of the sinus without puncturing it.

'I think we can call that a complete removal,' I said.

'I don't think I'm going to have time to do Mr Williams – the man with the frontal tumour,' Tim said. 'I've got a clinic starting at one. I'm terribly sorry. Could you possibly do him? And take out as much tumour as you can? Get him some extra time?'

'I suppose I'll have to,' I replied, disliking having to operate on patients I had not spoken to in detail myself, and not at all sure as to whether surgery was really in the patient's best interests.

So Tim went off to do his outpatient clinic and Samih finished the operation, filling the hole in the girl's skull with quick-setting plastic cement and stitching together the layers of her scalp. An hour later, Mr Williams was wheeled into the anaesthetic room next to the operating theatre. He was in his forties, I think, with a thin moustache and a pale, rather vague expression. He must have been quite tall as his feet, clad in regulation white anti-embolism stockings with the bare toes coming out at the ends, stuck out over the edge of the trolley.

'I'm Henry Marsh, the senior surgeon,' I said, looking down at him.

'Ah,' he said.

'I think Tim Jones has explained everything to you?' I asked.

It was a long time before he replied. It looked as though he had to think very deeply before replying.

'Yes.'

'Is there anything you would like to ask me?' I said.

He giggled and there was another long delay.

'No,' he eventually replied.

'Well, let's get on with it,' I said to the anaesthetist and left the room.

Samih was waiting for me in the operating theatre, beside the wall-mounted computer screens where we can look at our patients' brain scans. He already had Mr Williams's scan on the screens.

'What should we do?' I asked him.

'Well, Mr Marsh, it's too extensive to remove. All we can do is a biopsy, just take a small part of the tumour for diagnosis.'

'I agree, but what's the risk with a biopsy?'

'It can cause a haemorrhage, or infection.'

'Anything else?'

Samih hesitated, but I did not wait for him to reply.

I told him how if the brain is swollen and you only take a little bit of tumour out, you can make the swelling worse. The patient can die after the operation from 'coning': the swollen brain squeezes itself out of the confined space of the skull, part of it becoming cone-shaped where it is forced out of the skull through the hole at its base called the foramen magnum ('the big hole' in Latin), where the brain is joined to the spinal cord. This process is invariably fatal if it is not caught in time.

'We have to take enough tumour out to allow for any post-op swelling,' I said to Samih. 'Otherwise it's like kicking a hornet's nest. Anyway, Tim said he was going to remove as much of the tumour as possible as this might prolong his life a bit. What sort of incision do you want to make?'

We discussed the technicalities of how to open Mr Williams's head while waiting for the anaesthetists to finish anaesthetizing him, and to attach the necessary lines and tubes and monitors to his unconscious body.

'Get his head open,' I told Samih, 'and give me a shout when you've reached the brain. I'll be in the red leather sofa room.'

The scan had shown that the left frontal lobe of Mr Williams's brain was largely infiltrated by tumour, which appeared on the scan as a spreading white cloud in the grey of his brain. Tumours like this grow into the brain instead of displacing it, the tumour cells pushing into the brain's soft substance, weaving their way between the nerve fibres of the white matter and the brain cells of the grey matter. The brain can often go on working for a while even though the tumour cells are boring into it like deathwatch beetles in a timber building, but eventually, just as the building must collapse, so must the brain.

I lay on the red leather sofa in the neurosurgeons' sitting room, slightly anxious, as I always am when waiting to operate, longing to retire, to escape all the human misery that I have had to witness for so many years, and yet dreading my departure as well. I am starting all over again, I said to myself once more, but am running out of time. The phone rang and I was summoned back to the theatre.

Samih had made a neat left frontal craniotomy. Mr Williams's forehead had been scalped off his skull and was reflected forward with clips and sterile rubber bands. His brain, looking normal but a little 'full', as neurosurgeons describe a swollen brain, bulged gently out of the opening Samih had sawn in his skull.

'We can't miss it, can we?' I said to Samih. 'The tumour's so extensive. But the brain's a bit full – we'll have to take quite a lot out to tide him over the post-operative period. Where do you want to start?'

Samih pointed with his sucker to the centre of the exposed surface of brain.

'Middle frontal gyrus?' I asked. 'Well, maybe, but let's go and look at the scan.' We walked the ten feet across the room to the computer screens.

'Look, there's the sphenoid wing,' I said to Samih. 'We should go in just a little above it, but you'll have to go deeper into the brain than you think from the scan as his brain is bulging out a bit.'

We returned to the table and Samih burned a little line across Mr Williams's brain with the diathermy forceps – a pair of forceps with electrical tips that we use for cauterizing bleeding tissue.

'Let's bring in the scope,' I said, and once the nurses had positioned the microscope, Samih gently pushed downwards with sucker and diathermy.

'It looks normal, Mr Marsh,' Samih said, a little anxiously. Even though there are all manner of checks and cross-checks to make sure we have opened the correct side of the patient's head, I always experience a moment of complete panic at times like this, and have to quickly reassure myself that we are indeed operating on the correct side – in this case the left side – of Mr Williams's brain.

'Well, the trouble with low-grade tumours is that they can look and feel like normal brain. Let me take over.'

So I started to cautiously prod and poke the poor man's brain.

'Yes, it looks and feels entirely normal,' I said, feeling a little sick as I looked through the microscope at the smooth, unblemished white matter. 'But we've *got* to be in tumour – there's so much of it on the scan.'

'Of course we are, Mr Marsh,' Samih said respectfully. 'Would Stealth or a frozen section have helped?'

These are techniques that would have reassured me that I was in the right place. Rationally I knew that I had to be in tumour – at least in brain infiltrated by tumour – but the man's brain looked and felt so normal that I could not suppress the fear that some bizarre mistake had occurred. Perhaps the wrong name was on the brain scan, or it hadn't been a tumour in the first place and the problem had got better on its own since the brain scan had been done. The thought of removing normal brain – however unlikely – was terrifying.

'Well, you're probably right, but it's too late now and, having started, I can't stop,' I said to Samih. 'I'll have to remove a lot of normal-looking brain to stop him swelling and dying post-op.'

The brain becomes swollen with the least provocation, and Mr Williams's brain was already ominously enlarging and starting to bulge out of his opened skull. At the end of a craniotomy – the medical name for opening a person's head – the skull is closed with little metal screws and plates and the scalp stitched back together over it. The skull becomes once again a sealed box. If there is very severe post-operative swelling as a reaction to the surgery, the pressure inside the skull will become critically raised and the brain will, in effect, suffocate and the patient can die. Surgery, especially for tumours within the actual substance of the brain like Mr Williams's, where you cannot remove all of the tumour, will inevitably cause swelling, and it is always important to remove enough tumour – to create space within the skull to allow for the swelling. The pressure in the patient's head after the operation will then not become dangerously high. But you always worry that you might have removed too much tumour and that the patient will wake up damaged and worse than before the operation.

I can remember two cases – both young women – from the early years of my career where my inexperience made me too timid and I failed to remove enough tumour. They both died from post-operative brain swelling within twenty-four hours after surgery. I learnt to be braver with similar cases in future – in effect, to take greater risks when operating on such tumours, because the deaths of the two women had taught me that the risks of not removing sufficient tumour were even greater. And yet both the tumours were malignant and the patients had a grim future ahead of them, even if the operations were to have been successful. Looking back now after thirty years, having seen so many people die from malignant brain tumours since then, these two tragic cases do not seem quite as disastrous as they did at the time.

This is about as bad as it gets, I thought with disgust as I started to remove several cubic centimetres of Mr Williams's brain, the sucker slurping obscenely. What's the glory in this? This coarse and crude surgery. This evil tumour, changing this man's very nature, destroying both himself and his family. It's time to go.

As I watched my sucker down the microscope, controlled by my invisible hands, working on the poor man's brain, teasing and pulling out the tumour, I told myself that I wouldn't have panicked in the past. I would just have shrugged and got on with it. But now that my surgical career was coming to an end, I could feel the defensive psychological armour that I had worn for so many years starting to fall away, leaving me as naked as my patients. Bitter experience of similar cases to Mr Williams's told me that the best outcome for this man would be if the operation killed him – but I felt unable to let that happen. I knew of surgeons in the distant past who would have done just that, but we live in a different world now. At moments like this I hate my work. The physical nature of our thought, the incomprehensible unity of mind and brain, is no longer an awe-inspiring miracle but instead a cruel and obscene joke. I think of my father slowly dying from dementia and his brain scan, and I look at the age-wrinkled skin of my hands, which I can see even through the rubber of my surgical gloves.

As I worked the sucker, Mr Williams's brain started slowly to sink back into his skull.

'That's enough space now, Samih,' I said. 'Close please. I'll go and find his wife.'

Later in the day I went up to the ITU to see the post-operative patients. The young Romanian woman was well, though she looked pale and a little shaken. The nurse at the end of her bed glanced up from the mobile computer where she was inputting data and told me that everything was as it should be. Mr Williams was three beds further down the row of ITU patients. He was sitting upright, awake, looking straight ahead.

I sat by his bedside and asked him how he felt. He turned to look at me and said nothing for a while. It was hard to know if his mind was blank or whether he was struggling to organize the thoughts in his disrupted, infiltrated brain. It was hard even to know what 'he' had now become. Once I would have waited only a short time for an answer. Many of my patients have lost – sometimes permanently, sometimes transiently – language or the ability to think and there is a limit to how long you can put up with waiting. But on this occasion, perhaps because I knew that this would probably never happen again and perhaps also as a silent apology to all the patients I must have hurried by in the past, I sat quietly for what felt like a long time.

'Am I going to die?' he suddenly asked.

'No,' I said, alarmed at the way he seemed to know what was going on after all. 'And if you were I promise I would tell you. I always tell my patients the truth.'

He must have understood that because he laughed – an odd, inappropriate sort of laugh. No, you are not going to die just yet, I said to myself, it is going to be much worse than that. I sat beside him for a while longer but it seemed he had nothing further to say.

Samih was waiting for me as usual at 7.30 the next morning at the nurses' desk. He was a junior doctor in the traditional mould and could not bear to think that he might not be in the hospital when I was there. When I was a junior it was inconceivable that I might leave the building before my consultant, but in the new world of shift-working doctors the master-and-apprentice form of medical training has largely disappeared.

'She's in the interview room,' he said. We walked down the corridor and I sat down opposite Mrs Williams. I introduced myself.

'I'm sorry we haven't met before. Tim was going to do

the operation but I ended up doing it. I'm afraid this is not going to be good news. What did Tim tell you?'

As a doctor you get used to patients and their families looking so very intently at you as you talk that sometimes it feels as though nails are being driven into you, but Mrs Williams smiled sadly.

'That it was a tumour. That it couldn't all be removed. My husband was pretty bright, you know,' she added. 'You're not seeing him at his best.'

'In retrospect, looking back, when do you think things started to go wrong?' I asked gently.

'Two years ago,' she said immediately. 'It's a second marriage for both of us – we married seven years go. He was a lovely man, but two years ago he changed. He was no longer the man that I had married. He started playing strange, cruel tricks on me . . .'

I did not ask what these might have been.

'It became so bad,' she went on, 'that we had more or less decided to go our separate ways. And then the fits started . . .'

'Do you have children?' I asked.

'He has a daughter from his first marriage but we have no children from our marriage.'

'I'm afraid I have to tell you that treatment won't get him better,' I said, very slowly. 'We can't undo the personality change. All we can do is possibly prolong his life and he may yet live for years anyway, but he will slowly get worse.'

She looked at me with an expression of utter despair – she could not have helped but hope that the operation would undo the horrors of the past, that her nightmare would come to an end.

'I thought it was the marriage that had gone wrong,' she said. 'His family all blamed me.'

'It was the tumour,' I said.

'I realize that now,' she replied. 'I don't know what to think . . .'

We talked for a while longer. I explained that we would have to wait for the pathology report on what I had removed. I said it was just possible I might have to operate again if the analysis showed that I had missed the tumour. The only potential further treatment would be radiation and, as far as I could tell, this had no prospect of making him any better.

I left her in the little interview room with one of the nurses – most of my patients' families prefer, I think, to cry after I have left the room, but perhaps that is wishful thinking on my part – perhaps they would prefer me to stay.

Samih and I walked back down the corridor.

'Well,' I said, 'at least the marriage was coming to an end, so I suppose it's a bit easier for her, but how can anybody know how to deal with something like this?'

I thought of the end of my first marriage fifteen years earlier and how cruel and stupid my wife and I had been to each other. Neither of us had had frontal brain tumours, though I wonder what deep and unconscious processes might have been driving our behaviour. I look back with horror at how little attention I paid to my three children during that time. The psychiatrist I was seeing at the time told me to become more of an observer, but I simply could not detach myself from the raging intensity of my feelings at being forced to leave my own home, so much of which I had built with my own hands. I feel that I have learnt a certain amount of wisdom and self-control as a result of that terrible time, but also wonder whether it might in part be simply because the emotional circuits in my brain are slowing down with age.

I went to see Mr Williams. The nurses had told me, when I had come onto the ward, that he had tried to abscond during the night, and they had had to keep the ward door locked. It was a fine morning and low sunlight streamed into the ward through the east-facing windows, over the slate roofs of south London. I found him standing in front of the windows in his pyjamas. I noticed that they were decorated with teddy bears. His arms were stretched out on either side as though to welcome the morning sun.

'How are you?' I said, looking at his slightly swollen forehead and the neatly curved incision behind it across his shaven head.

He said nothing in reply and gave me a vague, cryptic smile, slowly lowered his arms and shook my hand politely without saying a word.

The pathology report came back two days later and confirmed that all the specimen I had sent was infiltrated by a slow-growing tumour. It was going to take a long time to find any kind of long-term placement for Mr Williams and it seemed unlikely he could be managed at home, so I told my juniors to send him back to the local hospital to which he had first gone after the epileptic fits had started. The doctors and nurses there would have to find a solution to the problem. The tumour was certainly going to prove fatal, but it was impossible to know whether this would be a matter of months or longer. When I went round the ward early next morning I saw that there was a different patient in his bed and Mr Williams had gone.

2

LONDON

I had decided to resign from my hospital in London in a fit of anger in June 2014, four months before I came across the lock-keeper's cottage. Three days after handing in my letter of resignation I was in Oxford, where I live with my wife Kate at weekends, running along the Thames towpath for my daily exercise. I was panic-stricken about what I would do with myself once I no longer had my work as a neurosurgeon to keep me busy and my mind off the future. It was in exactly the same place, on the same towpath, but walking, not running, many years earlier, in a much greater state of distress, that I had decided to abandon my degree in politics, philosophy and economics at Oxford University – much to my parents' distress and dismay when they got to hear of it.

While I ran beside the river, I suddenly remembered a young Nepali woman with a cyst in her spine that had been slowly paralysing her legs. I had operated on her two months previously. The cyst turned out to be cysticercosis, a worm infection common in impoverished countries like Nepal but almost unheard of in England. She had returned to the outpatient clinic a few days earlier to thank me for her recovery; like so many Nepalis, she had the most perfect, gentle manners. As I ran – it was late summer, the river level was low and the dark-green water of the Thames seemed to be almost motionless – I thought of her and then thought of Dev, Nepal's first and foremost neurosurgeon, more formally known as Professor Upendra Devkota. We had been friends and surgical trainees together in London thirty years ago.

'Ah!' I thought. 'Perhaps I can go to Nepal and work with Dev. And I will see the Himalayas.'

Both decisions, separated by forty-three years – to abandon my first degree and to resign from my hospital – had been provoked by women. The first was a much older woman, a family friend, with whom I was passionately and wholly inappropriately in love. Although twenty-one years old, I was immature and sexually entirely inexperienced, and had had a repressed and prudish upbringing. I can see now that she seduced me, although only with one passionate kiss – it never went beyond that. She burst into tears immediately afterwards. I think she had been attracted by my combination of intellectual precocity and awkwardness. Perhaps she thought that she could help me overcome the latter. She probably later felt ashamed, and perhaps embarrassed, by my passionate, poetic response – the poems now long forgotten and destroyed. She died many years ago, but my intense embarrassment about this episode is still with me now, even though the kiss resulted in my finding a sense of meaning and purpose to my life. I became a brain surgeon.

I was confused and ashamed by the pangs of my frustrated and absurd love, and overwhelmed by feelings of both love and rejection. I felt there were two armies fighting within my head and I wanted to kill myself to escape them. I tried to compromise by pushing my hand through a window in the flat where I had student digs beside the Thames in Oxford, but the glass would not break or, rather, a deeper part of my self showed a sensible caution.

Unable to translate my unhappiness into physical injury, I decided to run away. I made the decision while walking along the Thames towpath in the early hours of the morning of 18 September 1971, having fortunately failed to hurt myself. The towpath is narrow, in summer dry and grassy, in winter muddy and with many puddles. It passes through Oxford and past Port Meadow, the wide flood meadow to the north of the city. My childhood family home was a few hundred yards away. I might even have seen it as I walked miserably along the river – the area was deeply familiar. If I had gone a little further and followed a narrow cut, linking the river to the Oxford canal, I would have come across the lock-keeper's cottage, but I think I had already turned back by then, having made my decision. The old man, though young at that time, would already have been living there.

I abandoned my university degree for unrequited love, but it was also a rebellion against my well-meaning father, whose belief in the virtue of attending Oxford or Cambridge university was an article of faith. He had been an Oxford don before moving to London. He deserved better from me, but such rebellious behaviour is buried deep within the psyche of many young people; and my father, the kindest of men, but who had himself once rebelled against his own father, resigned himself to my decision. I left my predictable professional career path to work as a hospital theatre porter in a mining town north of Newcastle. I hoped that by seeing other people suffering with 'real', physical illness I would somehow cure myself. My subsequent life as a neurosurgeon was to teach me that the distinction between physical and psychological illness is false – at least, that illnesses of the mind are no less real than those of the body, and no less deserving of our help. A friend's father, John Maud, was the general surgeon in the hospital, and although he had never met me, at his daughter's request he got me a job in his operating theatre. I find it quite extraordinary that he did this, just as I find it remarkable that my Oxford college agreed that I could return after a year's truancy. It is impossible to know how my life would have developed without so much help and kindness from others.

It was my experience as a theatre porter, watching surgeons operate, that led me to become a surgeon. It was a decision that came quite suddenly to me, while talking to my sister Elisabeth – a nurse by training – as she did her family's ironing, when I returned to London for a weekend. I had gone to visit her to hold forth at great length about my unhappiness. It somehow became clear to me – I can't remember how – that the solution to my unhappiness was to study medicine and become a surgeon. Perhaps Elisabeth suggested it to me. I took the train back to Newcastle on the Sunday evening. As I sat in the carriage, seeing myself reflected in the dark glass of the window, I knew that I had now found a sense of purpose and meaning. It would be another nine years, however, when I was already a qualified doctor, before I discovered the all-consuming love of my life – the practice of neurosurgery. I have never regretted that decision, and have always felt deeply privileged be a doctor.

I am not sure, however, if I would take up medicine or neurosurgery now, if I could start my career all over again. So many things have changed. Many of the most challenging neurosurgical operations – such as operating on cerebral aneurysms – have become redundant. Doctors are now subject to a regulatory bureaucracy that simply did not exist forty years ago and which shows little understanding of the realities of medical practice. The National Health Service in

England – an institution I passionately believe in – is chronically starved of funds, since the government dares not admit to the electorate that they will need to pay more if they want first-class health care. Besides, there are other, more pressing problems now facing humanity than illness.

As I returned to Newcastle with my new-found sense of having a future, I read the first issue of a magazine called *The Ecologist*. It was full of gloomy predictions about what was going to happen to the planet as the human population continued to grow exponentially, and as I read it I wondered whether becoming a doctor, healing myself by healing others, might not be a little self-indulgent. There might be more important ways of trying to make the world a better place – admittedly less glamorous ones – than by being a surgeon. I have never entirely escaped the view that being a doctor is something of a moral luxury, by which doctors are easily corrupted. We can so easily end up complacent and self-important, feeling ourselves to be more important than our patients.

A few weeks later, back at work as a theatre technician, I watched a man undergoing surgery to his arm. He had deliberately pushed his hand through a window in a drunken rage and his hand had been left permanently paralysed by the broken glass.

The other woman who quite unintentionally played a pivotal role in my life – at the end of my neurosurgical career – was the medical director of my hospital. She was sent one day by the hospital's chief executive to talk to the consultant neurosurgeons. I believe that we had the reputation of being arrogant and uncooperative. We were too aloof and not playing our part. I was probably considered to be one of the worst offenders. She came into our surgeons' sitting room – the one with the red leather sofas that I had bought some years previously – accompanied by a colleague who was called, I think, the Service Delivery Unit Leader (or some similarly absurd title) for the neurosurgery and neurology departments. He was a good colleague and on several occasions had saved me from the consequences of some of my noisier outbursts. He was suitably solemn on this occasion, and the medical director was looking perhaps a little anxious at the prospect of disciplining eight consultant neurosurgeons. She sat down and carefully placed her large pink handbag beside her on the floor. Our Service Delivery Unit Leader made a little introductory speech and handed over to the medical director.

'You have not been following the Trust dress code,' she declared. Apparently this meant that the consultant neurosurgeons had been seen wearing suits and ties. I had always thought that dressing smartly was a sign of courtesy to my patients, but apparently it now posed a deadly risk of infection to them. A more probable, albeit unconscious, explanation for the ban – which came from high up the NHS hierarchy – was that the senior doctors should not look any different from the rest of the hospital staff. It's called teamwork.

'You have not been showing leadership to the juniors,' the medical director continued. This meant, she told us, that we had not been making sure that the junior doctors had been completing the Trust computer work on time when patients were discharged. In the past we had had our own neurosurgical discharge summaries, which had been exemplary, and I had always taken some pride in them, but they had now been replaced by a Trust-wide, computerized version of such appallingly poor quality that I, for one, had lost all interest in making sure that the juniors completed them.

'If you do not follow Trust policies, disciplinary action will be taken against you,' she concluded. There was no discussion, no attempt to persuade us. The problem, I knew, was that the hospital was about to be inspected by the Care Quality Commission, an organization that puts great store by dress policy and the completion of paperwork. She could have said that she knew this was all rather silly, but could we please help the hospital, and I am sure we would all have agreed – but no, it was to be disciplinary action. She picked up her pink handbag and left, followed by the Service Delivery Unit Leader, who looked a little embarrassed. So I sent off my letter of resignation the next day, unwilling to work any longer in an organization where senior managers could demonstrate such a lack of awareness of how to manage well, although I prudently postponed the date of my departure until my sixty-fifth birthday so that my pension would not suffer.

It is often said that it is better to leave too early rather than too late, whether it is your professional career, a party, or life itself. But the problem is to know when that might be. I knew that I was not yet ready to give up neurosurgery, even though I was so anxious to stop working in my hospital in London. I hoped to go on working part-time, mainly abroad. This would mean that I would need to be revalidated by the General Medical Council if I were to remain a licensed doctor.

Aircraft pilots need to have their competence reassessed every few years and, it is argued, it should be no different with doctors, because both pilots and doctors have other people's lives in their care. There is a new industry called Patient Safety, which tries to reduce the many errors that occur in hospitals and which are often responsible for patients coming to harm. Patient Safety is full of analogies with the aviation industry. Modern hospitals are highly complex places, and many things can go wrong. I accept the need for checklists and trying to instil a blame-free culture, so that mistakes and errors are identified and, hopefully, avoided. But surgery has little in common with flying an aircraft. Pilots do not need to decide what route to fly or whether the risks of the journey are worth taking, and then discuss these risks with their passengers. Passengers are not patients: they have chosen to fly, patients do not choose to be ill. Passengers will almost certainly survive the flight, whereas patients will often fail to leave the hospital alive. Passengers do not need constant reassurance and support (apart from the little charade where the stewardesses and stewards mime the putting-on of life jackets and point confusingly to the emergency exits). Nor are there anxious, demanding relatives to deal with. If the plane crashes, the pilot is usually killed. If an operation goes wrong, the surgeon survives, and must bear an often overwhelming feeling of guilt. The surgeon must shoulder the blame, despite all the talk about blame-free culture.

To revalidate doctors is important but not easy, and it took the General Medical Council in Britain many years to decide how to do it. As well as being 'appraised' by another doctor, I had to complete a '360-degree' assessment by several colleagues, and one by fifteen consecutive patients. I was tempted, when instructed to provide the names of colleagues, to name ten people who disliked me (alas, not very difficult), but I chickened out, and instead listed various people who would be unlikely to find great fault with me. They ticked the online boxes, saying how good I was, and how I achieved a satisfactory 'work–life balance', and I returned the favour when they sent me their 360-degree forms.

I was provided with fifteen questionnaires to hand out to patients. The exercise was managed by a private company – one of the many profitable businesses to which much NHS work is now outsourced. These companies prey off the NHS like hyenas off an elderly and disabled elephant – disabled by the lack of political will to keep it alive.

I was told to ask the patients to complete the lengthy, two-sided form after I had seen them in my outpatient clinic and to have them return the forms to me. Not surprisingly, I was on my best behaviour. Besides, the patients would probably have been reluctant to criticize me to my face. My patients obediently filled in the forms. It seemed to me that whoever would be examining them might well suspect that I had fraudulently completed them myself, as all the completed forms were both eulogistic and anonymous. I was tempted to do this but to accuse myself of being impatient and unsympathetic – in short, of being a typical surgeon – and see if this made any difference to the absurd charade.

My first neurosurgical post had been as a senior house officer in the hospital where I had trained as a medical student. There were two consultant neurosurgeons, the younger one very much my mentor and patron. The senior surgeon retired shortly after I started working in the department. He rang me once at night when I was on call, seeking advice about a friend of his who had passed out at home, asking whether this might be due to his blood-pressure drugs. It was fairly obvious that the friend was himself. I remember once standing with him in front of an X-ray screen looking at an angiogram – an X-ray that shows blood vessels – of a patient with a difficult aneurysm, and him telling me to ask his younger colleague to take over the case.

'By my age, aneurysm surgery is not good for the coronaries,' he said. I knew that recently one of the senior neurosurgeons in Glasgow had clipped an aneurysm and then immediately collapsed with a major heart attack.

My senior consultant's career ended gloriously with a

successful operation on a large benign brain tumour in a young girl. She recovered perfectly and a few days later, still in her hospital gown and with a shaven head, came to his retirement party to present him with a bouquet of flowers. I believe that he died a few months afterwards. My own surgical career, thirty-four years later, was to end ignominiously.

I had two weeks left before retiring and I was looking at a brain scan with my registrar, Samih.

'Fantastic case, Mr Marsh!' he said happily, but I did not reply. Until recently, I would have said exactly the same myself. The difficult and dangerous operations were always the most attractive and exciting ones, but as my career approached its end I was finding that my enthusiasm for such cases, and for the risk of disaster, was rapidly diminishing. The thought of the operation going badly, and of my leaving a wrecked patient behind me after my retirement, filled me with dismay. Besides, I thought, as I am soon to give all this up, why must I go on inflicting it on myself? But the patient had been referred to me personally by one of the senior neurologists. Suggesting that one of my colleagues do the operation instead was out of the question: it was just not compatible with my self-esteem as a surgeon.

'It should separate away from all the vital bits,' I said to Samih, pointing to the tumour on the scan. The tumour was growing at the edge of the foramen magnum. Damage to the brainstem or the nerves branching off it can be catastrophic for the patient, including paralysis of swallowing and coughing. This can lead to fluid in the mouth getting into the lungs and causing a very severe form of pneumonia that can easily be fatal. At least the tumour appeared benign. It did not look as though it would be stuck to the brainstem and spinal nerves so, at least in theory, it should be possible to remove the tumour without causing severe damage. But you can never be certain.

It was Sunday evening and Samih and I were sitting in front of the computer at the nurses' station on the men's ward. We both regretted the fact that our work together was soon to end. The close relationship you can have with your trainees is one of the great pleasures of a surgeon's life.

It was early March, and it was dark outside but the sky was clear; there was a very bright full moon, low over south London, which I could see through the ward's long line of windows. There had been a scent of spring in the air as I had bicycled in to work, along the back streets, the moon cheerfully racing along beside me over the slate roofs of the terraced houses.

'I haven't met him yet,' I said. 'So we had better go and talk to him.'

We found the patient in one of the six-bed bays, the curtains drawn around the bed.

'Knock, knock,' I said, drawing the curtain aside.

Peter was sitting up. There was a young woman in the chair beside the bed. I introduced myself.

'I'm so pleased to see you at last,' he said, looking much happier than most of my patients when I first meet them. 'The headaches have really been getting awful.'

'Have you seen the scan?' I asked.

'Yes, Dr Isaacs showed it to me. The tumour looked huge.'

'It's not that big,' I replied. 'I have seen many bigger, but then one's own tumour always looks enormous.'

Samih had pulled along one of the new mobile computer stations from the corridor and placed it at the end of Peter's bed. He summoned up the brain scan while we talked.

'That's a centrimetric scale there,' I explained, pointing to the edge of the scan. 'Your tumour is four centimetres in diameter. It's causing hydrocephalus – water on the brain – it's acting like a cork in a bottle and trapping the spinal fluid in your head where it is supposed to drain out at the bottom of your skull. Without treatment – although I apologize for terrorizing you – you only have a few weeks to live.'

'I can believe that,' he said. 'I've been feeling really lousy, though the steroids Dr Isaacs started me on helped a bit.'

We talked for a while about the risks of the operation – death or a major stroke were possible but unlikely, I said, and he might have difficulty swallowing. He nodded and told me that in recent weeks he had sometimes choked when eating. We talked also about his work, and about his children. I asked his wife what they knew about their father's illness.

'They're only six and eight,' she said. 'They know their Daddy is coming to hospital and that you are going to make his headaches better.'

While we talked, Samih filled up the long consent form and Peter signed it quickly.

'I'm not at all frightened,' he said, 'and I'm really glad I've got you to do it just before you retire.' I let this pass – patients want to think their surgeon is the best and don't particularly like it when I tell them that I am not and that I am dispensable. Samih noted his wife's phone number down on the edge of the consent form.

'I'll ring you after the op,' I said to Peter's wife. 'See you tomorrow.' I waved to Peter and slipped out between the curtains. There were five other men in the room who looked up at me as I left – no doubt they had all listened to the conversation with great interest.

As I cycled to work next morning, I reflected on the strange fact that almost forty years of working as a surgeon were coming to an end. I would no longer have to feel constantly anxious, with my patients so often on the edge of disaster, yet for almost forty years I had never had to worry about what to do each day. I had always loved my work, even though it was often so painful. Every day was interesting; I loved looking after patients, I loved the fact that I was – at least in my own little hospital pond – quite important, indeed my work had frequently felt more like a glorious opportunity for adventure and self-expression than mere work. It had always felt profoundly meaningful. But in recent years this love had started to fade. I attributed this to the way in which working as a doctor felt increasingly like being an unimportant employee in a huge corporation. The feeling that there was something special about being a doctor had disappeared – it was just another job, I was just a member of a team, many of whose members I did not even know. I had less and less authority. I felt less and less trusted. I had to spend more and more time at meetings stipulated by the latest government edicts that I felt were often of little benefit to patients. We spent more time talking about work rather than actually working. We would often look at brain scans and decide whether the patient should be treated or not without any of us having ever seen the patient. Like almost all the doctors I knew, I was becoming deeply frustrated and alienated.

And yet despite this, I was still burdened with an overwhelming sense of personal responsibility for my poor patients. But perhaps my discontent was because I had less and less operating to do – although I was lucky compared to many other surgeons in that I still had two days of operating a week. Many of my colleagues are now reduced to a single day each week; you may well wonder what they are supposed to do for the rest of the week. Recent increases in the number of surgeons have not been matched by any increase in the facilities we need in order to operate. Or then again, perhaps it was simply because I was getting old and tired and it was time to go. Part of me longed to leave, to be free from anxiety, to be master of my own time, but another part of me saw retirement as a frightening void, little different from the death, preceded by the disability of old age and possibly dementia, with which it would conclude.

There had been fewer emergency admissions than usual over the course of the weekend and there were empty beds on the ITU, so I was told that my list could start on time. The anaesthetist, Heidi, had been away on prolonged leave to look after her young son and was now back at work part-time. We were old friends and I was relieved to see her. The relationship between anaesthetist and surgeon is critical, especially if there is going to be trouble, and having colleagues who are friends is all-important. I walked into the anaesthetic room where Heidi and her assistants had Peter already asleep. The ODA – the operating department assistant, whose job is to help the anaesthetist – was stretching a wide band of Elastoplast across his face to keep the endotracheal tube – the tube which Heidi had inserted through his mouth, down his throat and into his lungs – in place. His face now disappeared beneath the Elastoplast, and the process of depersonalization that starts as the intravenous anaesthetic takes effect and the patient becomes unconscious was now complete.

I have watched that process thousands of times – it is, of course, one of the miracles of modern medicine. One moment the patient is talking, wide awake and anxious – although a good anaesthetist like Heidi will be soothing and reassuring – and the next instant, as the intravenously injected drug travels up the veins of the arm via the heart to the brain, the patient sighs, the head falls back a little, and he or she is suddenly and deeply unconscious. As I watch, it still looks to me as though the patient's soul is leaving the body to go I know not where and all I now see is an empty body.

'It might bleed a bit,' I said to Heidi, 'and the brainstem might be a problem.' Sudden and alarming changes in the patient's heart rate and blood pressure, even cardiac arrest, can occur if you get into trouble with the lower part of the brainstem, known as the medulla oblongata.

'Not to worry,' said Heidi. 'We're prepared. Big IV and plenty of blood cross-matched, ready in the fridge.'

Peter was wheeled into the operating theatre and, having assembled the theatre staff, we rolled him off the trolley face-down onto the operating table with Samih holding his head.

'Prone, neutral position, head well flexed,' I told him. 'Get him in the pins. Midline incision with the craniectomy more to the left and take out the back of C1. Give me a shout when you've done that and you're down to the dura and I'll come and join you.'

I left the operating theatre and went round to the surgeons' sitting room for the regular Monday morning meeting with my consultant colleagues. The meeting had already started, with our two line managers in attendance – both of whom, I might add, I liked and got on well with. The meetings were to discuss the day-to-day business of the neurosurgical department and the managers would sometimes tell us about the department's 'financial position'. Much of the meeting was spent letting off steam about all the petty frustrations and inefficiencies of working in a large hospital. There was a sky-blue cushion in the shape of a brain that had been given to me by the sister of one of my American trainees and sometimes we would throw it around the room as we talked, rather like holding the conch shell in Golding's

Lord of the Flies. Sean, the senior of the two managers, was talking. He declined to hold the cushion when I threw it at him.

'I'm afraid that this last year we made only one million pounds' profit for the Trust whereas the year before we made four million, even though we did not do any more work. We used to be one of the most profitable departments in the Trust but that is no longer the case.'

'But where on earth did the three million go?' somebody asked.

'It's not very clear,' Sean replied. 'We spent a lot on agency nurses. And you're spending a lot more on putting metalwork into people's spines and you're doing too many emergencies – we get only thirty per cent payment if you exceed the target for emergency work.'

'It's so bloody ridiculous,' I snorted. 'What would the public say if they knew we got penalized for saving too many lives?'

'You know the reason,' Sean said. 'It's to stop hospitals making cases into emergencies when they're not emergencies and over-claiming.'

'Well, we never did that,' I replied.

I should explain that 'profit' in an NHS department is not profit in the usual sense – instead it is whether we have exceeded our 'financial target' or not, which is based on previous performance and is an arcane process that I find entirely incomprehensible. Any 'profit' that we make goes to prop up less profitable parts of the Trust, so, despite the introduction into the NHS of the incentives and penalties so loved by economists, there is little real motive at a clinical level, on the shop floor, to work more efficiently. Besides, whenever there does seem to be any extra money, it all appears to be spent on employing more and more members of staff, as though to encourage the existing members of staff to do less work.

The conversation meandered on for a while, discussing the problem of spinal implants. There is no easy answer to this question. As intracranial neurosurgery has declined, replaced by non-surgical methods such as the radiological treatment of aneurysms and highly focused radiation for tumours, neurosurgeons (and there are ever-increasing numbers of them, all keen to operate) have moved into spinal surgery. This is largely about inserting all manner of very expensive titanium nuts and bolts and bars into people's backs, for cancer or for backache, although the evidence base and justification for such surgery, at least for back pain, are very weak. Even with the cancer patients – metastatic cancer often spreads to the spine – it can be a moot point as to whether to operate or not as the poor patient is going to die anyway, sooner or later, from the underlying cancer. Spinal implant surgery is major surgery and is a six-billion-dollar-a-year business in the US. It is a prime example of the 'over-treatment' that is a growing problem in modern health care, and especially in commercial, marketized health-care systems such as in America.

I stopped doing such surgery myself some years ago in order to concentrate on brain surgery, so I was happy to abandon the conversation when I was summoned back to the operating theatre, where Samih had started the operation.

'Let's have a look,' I said, and I leant forward, taking care not to touch the sterile drapes, to peer into the large hole in the back of Peter's head. 'Very good,' I commented. 'Open the dura and I'll go and put some gloves on. Jinja,' I said to the circulating nurse (the nurse who is not scrubbed up and does the fetching and carrying while the operation proceeds), 'can you get the scope in please?'

While Jinja shoved the heavy scope up to the operating table I scrubbed up at the large sink in the corner of the room – a soothing and deeply familiar act, although always accompanied by a feeling of tension in the pit of my stomach. I must have done this many thousands of times over the years and yet now I knew that it was soon to end – at least in my home country.

Jinja came and tied up the laces at the back of my blue gown and I marched up to the table where Peter lay hidden under the sterile blue drapes, with only the gaping and bloody hole in the back of his head to be seen, brilliantly lit by the operating lights. Samih opened the dura – the leathery, outer layer of the meninges – with a small pair of scissors while I watched. I then took over. I sat down on the operating chair with its arm rests. The first rule of microscopic surgery, I tell my trainees, is to be comfortable, and I usually sit when operating, although in some departments this is not considered to be very manly, and the surgeons stand throughout the procedure, often for very many hours on end.

It was easy enough to find the tumour – a bright-red ball shining in the microscope's light – a few millimetres beneath the back part of the brain, the cerebellum. To the left would be the all-important brainstem, and to the right and deep down the lower cranial nerves, scarcely thicker than thread; but all this was hidden by the tumour. I would not be able to see them until the very end of the operation, when I had removed most of the tumour. As soon as I touched the tumour with the sucker, blood spurted up out of it.

'Heidi,' I said, 'it's going to bleed.'

'No problem,' came the encouraging reply, and I settled down to attack the tumour.

'If the blood loss gets too much,' I said to Samih, 'your anaesthetist might ask you to stop and pack the wound, but then you worry you might damage the brain with the packing. If it looks as though the patient is going to bleed to death – to exsanguinate – sometimes you just have to operate as quickly as possible, get the tumour out before the patient dies and just hope you haven't damaged anything. The bleeding usually stops once the tumour is all out.'

'I saw you do a case like that when you came to Khartoum,' Samih commented.

'Ah yes. I'd forgotten that. He did OK though . . .'

It took four hours of intense concentration to get the tumour out. Down the three-centimetre-wide hole in Peter's head, all I could see was bright-red arterial blood, welling endlessly upwards. There was no way I could see the brain and no way I could delicately dissect the tumour off it. To my disappointment I did not enjoy the operation, which I think I would have done in the past. I should have arranged to do the operation jointly, I told myself, with a colleague. This greatly reduces the stress of operating, but I had not expected the tumour to bleed quite so much, and it is always difficult as a surgeon to ask for help, as bravery and self-reliance are seen as such an important part of the job. I would hate my colleagues to think that I was getting old and losing my nerve.

'Look, Samih,' I said, 'the damn thing did separate away.' With the tumour finally removed and the bleeding stopped, we could see the brainstem, and the lower cranial nerves and the vertebral artery all perfectly preserved. It made me think of the moon, appearing from behind clouds and transforming the night. It was a good sight.

'We were lucky,' I said.

'No, no,' said Samih, obeying the first rule for all surgical trainees, flattering me. 'That was fantastic.'

'Well, it didn't feel it,' I replied, and then shouted across to the far end of the table, 'Heidi, what was the blood loss?'

'Only a litre,' she said happily 'No need to transfuse him. His haemoglobin is still one hundred and twenty.'

'Really? It felt like a lot more,' I said, thinking that maybe I had been unnecessarily nervous during the operation. I consoled myself with the thought that perhaps all the years of experience counted for something after all. But Peter was going to be all right and that was all that mattered, and his young children would be happy that I had cured their Daddy's headaches.

'Come on, Samih,' I said, 'let's close.'

Peter awoke well from the operation. His voice was hoarse but I checked that he could cough, so I was not worried that he was at risk of aspiration.

I went back to the hospital late in the evening to see the post-op cases. I went in most evenings: I live nearby so it was easy for me and I knew that my patients liked seeing me on the evenings both before and after their surgery. It was also a private protest against the way in which doctors now are expected to work shifts with fixed hours, and medicine is no longer perceived as a vocation, a true profession. Many doctors now seem to have the same expectation.

I walked onto the ITU and found Peter among the two long lines of beds on either side of the warehouse-like room, each with its own nurse at the foot end, and a little forest of high-tech monitoring equipment at the head end.

'How is he?' I asked the nurse.

'He's OK,' came the reply. There are so many ITU nurses that I know only a few of them and I did not recognize this one. 'We had to put a nasogastric down in case he aspirated . . .'

To my surprise, when I looked at Peter, who was sitting upright in his bed, wide awake, I saw that somebody had indeed put a nasogastric tube up his nose and then taped it to his face. I was angry that he had been subjected to the unpleasant procedure of having the tube inserted; it should not have been done, as he did not need it. The tube is pushed up the nose and then down the back of the throat into the stomach – a very unpleasant experience, I am reliably informed by my wife Kate, who has personal experience of it. Nor is it an entirely harmless procedure, and cases have been recorded of the end of the tube getting into the lungs and causing aspiration pneumonia and death, or even getting into the brain. These are, admittedly, rare complications, but after such a difficult but successful operation I was furious that it had been done. The decision to insert it had been made by one of the ITU doctors, clearly less experienced than I was, and the doctor on duty on the ITU for the night denied all knowledge of it. There seemed little point in blaming the nurse. I asked Peter how he felt.

'Better than I expected,' he said in a slightly hoarse voice, and then proceeded to thank me again and again for the operation. I bid him goodnight and told him that we'd remove the wretched nasogastric tube in the morning.

I went into work next morning, and immediately went with Samih to the ITU. There was a different nurse at the end of Peter's bed whom, once again, I did not recognize. Peter was awake and told me that he'd managed to sleep a little – quite an achievement in all the inhuman noise and bright lights of the ITU. I turned to the nurse.

'I know you didn't insert the nasogastric tube, but please take it out,' I said.

'I'm sorry, Mr Marsh, but he will have to be checked by SALT.'

SALT are the speech and language therapists who some years ago started to assume responsibility for patients with swallowing problems as well as speech difficulties. I had had several disagreements with speech therapists in the past when they had refused to sanction removal of nasogastric tubes which in my opinion the patients did not need. As a result several patients had been kept in hospital being unnecessarily tube-fed, despite my protests. I was not the speech therapists' favourite neurosurgeon.

'Take the tube out,' I said, between gritted teeth. 'It should never have been inserted in the first place.'

'I'm sorry Mr Marsh,' the nurse replied politely, 'but I won't.'

I was seized by a furious wave of anger.

'He doesn't need the tube!' I shouted. 'I will take responsibility. It is perfectly safe. I did the operation – the brainstem and cranial nerves were perfectly intact at the end, he's got a good cough . . . take the bloody tube out.'

'I'm sorry Mr Marsh,' the hapless nurse began again. Overcome with rage and almost completely out of control, I pushed my face in front of his, took his nose between my thumb and index finger and tweaked it angrily.

'I hate your guts,' I shouted, turning away, impotent, furious and defeated, to wash my hands at the nearest sink. We are supposed to clean our hands after touching patients, so I suppose the same applies to assaulting members of staff. Years of frustration and dismay at my steady loss of authority, at the erosion of trust and the sad decline of the medical profession, had suddenly exploded – I suppose because I knew I was to retire in two weeks' time and suddenly could no longer restrain my rage and feeling of intense humiliation. I stormed off the ward followed by Samih, leaving a little group of amazed nurses standing at the end of Peter's bed. I do not often lose my temper at work and have certainly never laid a hand on a colleague before.

I slowly calmed down and returned later in the day to the ITU to apologize to the nurse.

'I'm very sorry,' I said. 'I shouldn't have done that.'

'Well, what's done is done,' he replied, though I did not know what he meant and wondered whether he would be making an official complaint – to which I felt he was fully entitled. Towards the end of the day I received an email from the matron for the ITU saying that she had learnt that there had been an 'incident' on the ITU and asking me to come and talk to her the next day.

I went home in a state of craven and abject panic, the like of which I had scarcely ever known before. It took me a long time to calm down – I was so pathetically frightened by the prospect of some kind of official disciplinary action being launched against me. Where's the brave surgeon now? I asked myself as I lay on my bed, shaking with fear and anger. It's time to go, it really is.

Next morning I duly reported to the ITU matron – a colleague I knew well and had been working with for many years. It brought back memories of being summoned to the headmaster's office at school for some misdemeanour in the past, and of my intense anxiety as I waited outside the door. Sarah, the ITU matron, and I had been together at the old hospital which had been closed twelve years earlier. It had become something of an anomaly: a single-specialty hospital, with a staff of about 180, dealing only with neurosurgery and neurology in a garden suburb surrounded by trees and gardens. There were some good clinical reasons for integrating us into the major hospital where we now work, with a staff of many thousands; and the site of the old hospital, Atkinson Morley's in Wimbledon (AMH), was of course far too beautiful to be a mere hospital. It was sold for commercial development and the hospital turned into apartments that now cost millions of pounds.

But we lost a lot as well – above all the friendly working relationships that can come when you work in a small organization where everybody knows each other on a personal level and work together on the basis of personal obligation and friendship. The efficiency of the hospital was a perfect illustration of Dunbar's number – that magic number of 150. The size of our brain, Robin Dunbar, an eminent evolutionary anthropologist at Oxford University, has argued (and the brain size of other primates), is determined by the size of our 'natural' social group, when humans and their brains evolved in small hunting and gathering groups. We have the largest brains among primates, and the largest social group. We can relate to about 150 people on an informal, personal basis, but beyond that leadership, impersonal rules and job descriptions become necessary.

So Sarah knew me quite well. Some of the comradely atmosphere of the old hospital had been preserved, despite the best efforts of the management to merge our department into the anonymous collective of the huge hospital where we now worked. I think anybody else in the nursing hierarchy of the hospital would have initiated some kind of formal disciplinary procedure against me.

'I'm ashamed of myself,' I told her. 'I suppose it happened partly because I know I'm leaving . . .'

'Well, he wasn't to know that SALT for you is like a red rag to a bull. He doesn't want to make a formal complaint but he said you were very frightening and it brought back memories of an assault he suffered some months ago.'

I hung my head in shame and remembered how my first wife had told me how terrifying I could look, as our marriage fell apart with furious arguments.

'He handled me very well and kept admirably calm,' I

said. 'Please thank him when you next see him. It won't happen again,' I added with a slight smile. Sarah knew well enough that I was about to retire. I left her office and went round to the men's ward where Peter had been sent the previous evening. At least the senior nurse there had been happy to remove the wretched nasogastric tube at my request and it was nice to find Peter drinking a cup of tea without any problems, although he certainly had a very hoarse voice.

'I'm not supposed to attack the nurses in front of the patients,' I said. 'I'm really sorry.'

'No, no, not at all,' he replied with a croaking laugh. 'I told them I didn't need the tube and could swallow perfectly well but they wouldn't listen to me and just shoved it in. I was on your side.'

My last operation here, I thought, as I cycled home in the evening.

I finally left my hospital two weeks later, having cleared my office. I disposed of the accumulated clutter that a consultant surgeon acquires over the course of his career. There were letters and photographs from grateful patients, presents and plaques, and outdated textbooks, some of which had belonged to the surgeon whom I had replaced almost thirty years earlier. There were even some books, and an ophthalmoscope, that had belonged to his predecessor, the famous knighted surgeon who seventy years ago had created the neurosurgical department in which we worked. I spent days emptying eight filing cabinets, occasionally stopping to read with amusement some of the pronouncements and plans and protocols, reports and reviews, generated by a labyrinth of government offices and organizations, mostly now defunct, renamed, reorganized or restructured. And there were files dealing with cases where I had been sued, or bitter letters of complaint, from which I quickly averted my eyes – the memory was so painful. Having done all this I left my office, empty, for my successor. I had no regrets whatsoever.

3

NEPAL

There was a minor earthquake in the evening, small enough to be exciting rather than frightening. We were sitting in the garden, in the dusk, the crescent moon in the west blood-red with the city's polluted air, when there was a sudden low sound, almost like a breath of wind or a subterranean thought – a fleeting presence of something of immense size and distance. The bench I was sitting on in the garden briefly shook as though somebody had nudged it, and thousands of voices rose up all around us in the night from the dark valley below, wailing, crying out in fear like the damned on hearing that they are to go down to hell, and all the dogs of Kathmandu started barking furiously. And then, when it became clear there was not to be a major quake like the one which had killed thousands of people the year before, everything fell quiet and we could hear the cicadas again.

I slept very well that night and woke to the dawn chorus of the birds singing in the garden. A pair of syncopated cuckoos were calling, while the hooded crows croaked and quarrelled in the camphor tree and all the cocks in the valley crowed. At ten past eight I set off for the hospital – it is a walk of which I never tire and, for reasons I struggle to understand, I feel more deeply content as I go to work each morning than I have ever felt before. The rising sun casts long and peaceful shadows. The air is often hazy with pollution, but sometimes I am lucky and I can see the foothills that surround the city and, just peaking above them in the distance, the snow-covered summit of Mount Ganesh, the elephant god.

At first I walk in silence, apart from the birdsong, past houses with cascades of crimson and magenta bougainvillea at the entrance, and Buddhist prayer flags, like coloured handkerchiefs on a washing line, on the roof. The houses are all built of rendered brick and concrete, painted cheerful colours and look like stacked-up matchboxes with balconies and roof terraces and the occasional added gable or Corinthian column. Sometimes there is a peasant woman watching over a couple of cows, peacefully grazing on the thin and scruffy grass at the side of the cracked, uneven road. There is rubbish everywhere, and stinking open drains. Dogs lie sleeping on the road, probably worn out by a night's barking. Sometimes I walk past women carrying huge baskets of bricks on their backs, supported by straps across their foreheads, to a nearby building site. After the houses there are then many small shops, all open at the front; looking into them is like opening a storybook, or peeking into a doll's house.

Life here is lived on the street. There is the barber shaving a man with a cut-throat razor; another customer reads a newspaper while waiting, and the meat shop with ragged lumps of fresh meat and the severed head of a mournful, lop-eared goat looking blankly at me as I pass. There is the cobbler sitting cross-legged on the ground while he cuts soles from rubber sheeting, with cans of adhesive stacked against the wall. Cobblers are *dalits*, the untouchables of the Hindu caste system, and second only to the sweepers and cleaners, who are at the very bottom of society. He once repaired my brogue boots which have accompanied me all over the world, and which I polish assiduously every morning – the only practical activity I have when in Nepal, other than operating. He did a very good job of it and it was only when I later learned that he was a *dalit* that I understood why he at first looked awkward and embarrassed when I politely greeted him each morning as I passed his open workshop. There is the metalworker welding metal in a shower of blue sparks and a seamstress, with clothes hanging up at the front of her shop, while she sits at the back. I can hear the whirring of her sewing machine as I walk past. Motorbikes wind their way between the children in smart uniforms going to school. The children will look slightly askance at me – this is not a part of town to which expats normally come – and if I smile at them they give me a happy smile in return and wish me a good morning. I would not dare to smile at children back in England. There is a rawness, a directness to life here, with intense and brilliant colours, which was lost in wealthy countries a long time ago.

I walk past all these familiar sights to reach the main road, a melee of cars, trucks and pedestrians, with swarms of motorbikes weaving their way between them in a cloud of pollution, all blowing their horns. The broken gutters are full of rubbish, and next to them there are fruit vendors selling apples and oranges from mobile stalls that rest on bicycle wheels. There are long lines of colourful, ramshackle shops, and everywhere you look, hundreds of people going about their daily business, many of them wearing face masks which are, of course, useless against vehicle fumes. Electric cables droop like tangled black cobwebs from the pylons, which lean at drunken angles, and there are often broken ends with exposed wires, hanging down onto the pavement. I cannot even begin to imagine how any repair work is ever carried out. The women, with their fine faces, their jet-black hair swept back from their foreheads and their spectacularly colourful dresses and gold jewellery, transform what would otherwise often be depressing scenes of grinding poverty.

I have to cross the road to reach the hospital. I found this at first an unnerving experience. The traffic is chaotic and if you wait for a break in it, you will be there for a very long time. You must calmly step out onto the road, join the traffic, and walk slowly and predictably across, trusting the buses, vans and motorcycles to weave their way around you. Some of the motorcyclists have their helmets pushed back over their heads, so they look like the ancient Greek warriors to be seen on Attic vases. If you break into a run they are more likely to hit you by mistake. My guidebook to Nepal helpfully told me that 40 per cent of victims of road traffic accidents – RTAs, as they are called in the trade – are pedestrians. We admitted such cases every day to the hospital. I was to witness several fatal accidents. On one such occasion I passed a dead pedestrian on the Kathmandu ring road. He was sprawled on his face across the gutter, his legs bent out akimbo at an improbable angle like a frog's, with a group of curious onlookers watching silently as the police made notes. I have come to enjoy crossing the road – there is a feeling of achievement each time I get across it safely.

When I was a student almost fifty years ago, Kathmandu had been the fabled, near-mystical destination for many of my contemporaries. This was partly because cannabis grows wild in Nepal – and still does on building plots and derelict land in the city – but also because it was a place of pristine beauty and still living a life of medieval simplicity. They would trek overland. The world was a different place: you could safely travel through Syria, Iran and Afghanistan. Since then Kathmandu has also changed, almost beyond recognition. The population of Kathmandu has gone from a few hundred thousand twenty years ago to two and half million and it is the fastest-growing city in South East Asia. The new suburbs are entirely unplanned, without any proper infrastructure, occasionally with a few pathetic scraps of rice paddy or wheat field left as an afterthought between the cheap concrete buildings. There are open drains and dirt tracks, with rubbish and building materials strewn about everywhere. The roads are chaotic and the air is dark with pollution. You can rarely, if ever, see the high Himalayas to the north.

Nepal is one of the poorest countries in the world, shattered by a recent earthquake and with the ever-present threat of another catastrophic one to come. There are minor tremors every week. I am working with patients with whom I have only the most minimal human contact. The work is neurosurgical, so there are constant failures and disasters, and the patients' illnesses are usually more advanced and severe than in the West. The suffering of the patients and their families is often terrible, and you have to fight not to become inured and indifferent to so much tragedy. I can rarely, if ever, feel pleased with myself. The work, if I care to think about it, is often deeply upsetting and, compared to Public Health, of dubious value in a country as poor as Nepal. The young doctors I am trying to train are so painfully polite that I am never sure what they really think. I do not know whether they understand the burden of responsibility that awaits them if they ever become independent neurosurgeons. Nor do I know what they feel about their patients, or how much they care for them, as their English is limited and I cannot speak Nepali. What I do know is that most of them want to leave Nepal if they possibly can. Their pay and professional prospects here are poor compared to what they can find in wealthier countries. It is a tragedy affecting many low-income countries such as Nepal and Ukraine – the educated younger generation, the countries' future, all want to leave. I am working in a very alien, deeply superstitious culture with a cult of animal sacrifice, centred on blood.

Few, if any, of the patients and their families understand the unique and overarching importance of the brain, of the physical nature of thought and feeling, or of the finality of death. Few of the patients or their families speak English, and I feel very remote from them. They have wholly unrealistic expectations of what medicine can achieve, and take it very ill if things go badly, although they think we are gods if we succeed. I lead a life of embarrassing luxury compared to most people here – in my colleague Dev's guest house, with its little paradise of a garden – but I live out of a suitcase, with none of the property and possessions that dominate my life back in England. I am in bed by nine in the evening and up by five, and spend ten hours a day in the hospital, six days a week. I miss my home and family and friends intensely. Yet when I am here I feel that I have been granted a reprieve, that I am in remission, with the future postponed.

The day before my flight to Nepal had not been uneventful. I had reported to the private hospital where, for many years, I had worked in my own time, in addition to my work for the NHS, although I had stopped all private practice two years earlier. Over the preceding weeks I had noticed a slightly scaly lump growing on my forehead. One of the privileges of being a doctor is that you know to whom to go if you have a problem, and a plastic surgeon I knew well, and greatly liked, had told me the lump should be removed.

'You must have got the supra-orbital nerve. I can't feel a thing. The top of my head feels like wood,' I said to David once he had started, although I could feel the pressure of the scalpel cutting into my forehead. I had often subjected my own patients to this – although usually with much longer incisions and more local anaesthetic. This had been in order to saw into their skulls and expose their brains for an awake craniotomy, an operation I had pioneered for brain tumours, where you operate on the patient's exposed brain while they are awake. This was the first time that I could understand a little of what they would have experienced. I could feel David mopping up my blood as it ran down into my ear.

'Hmm,' he said. 'There are two points to it. It looks a bit invasive. You may need wider resection and skin-grafting.'

I felt a sudden surge of anxiety: although he was avoiding the word, he was obviously talking about cancer. I had thought removal of the little lump growing on my forehead was all going to be very simple. I now imagined myself with a large and ugly skin graft on my forehead. Perhaps I would need radiotherapy as well. I couldn't help but remember some of the patients I had treated with malignant scalp tumours that had eventually eaten their way through their skulls and bored into their brains.

'But it is curable, isn't it? And they don't normally metastasize do they?'

'Henry, it will all be fine,' David said reassuringly, probably amused by my anxiety.

'And can it wait two months?' I asked.

'Yes, I'm sure it can, but we'll have to see what the microscopy shows. How invasive it is. I'll email you.'

Doctors traditionally pay their colleagues for their services in wine, and before I left I arranged for some to be sent to David. Many years ago I operated on a local GP's wife with a difficult cerebral aneurysm, and she died immediately afterwards; I felt I was to blame. I was deeply ashamed when he sent me a case of wine some weeks after her death but it was, I now understand, an act of great professional kindness.

So I was on the plane to New Delhi next day, en route to Kathmandu, sporting a large, sticky plaster on the right side of my forehead, which I inspected gloomily in the mirror whenever I went to the cramped little toilet on the eight-hour flight, cursing my prostatism and skin cancer.

Having braved the traffic, I walk down the steep drive to Neuro Hospital, as it is called, set in a small valley off the main road. When the hospital was built ten years ago this was a rural area of paddy fields, but now it is almost entirely built up, although there is still one small paddy field left stranded, with a banana tree, next to the hospital.

The full name of the hospital Dev built is the National Institute for Neurology and Allied Sciences. It is large and spacious and spotlessly clean, with good natural light almost everywhere. The hospital is surrounded by gardens, just like AMH, the old hospital in Wimbledon, where Dev and I had trained together many years ago. Many of the patients – the women in brilliantly coloured dresses, deep reds, blues and greens, often with gold decorations – wait on the benches in front of the entrance. Dev planted a magnolia tree there, in memory of the magnolia tree that grew in front of AMH (that particular tree has now been felled as part of the conversion of the old and famous hospital into luxury flats). At night there will be many families sleeping on mats outside the side entrance. It is strange to come to a country as poor as Nepal and find such a sympathetic hospital, with so many windows and so much space, and so clean and well cared for. It incorporates all the lessons Dev learnt from working in small, specialist hospitals in Britain. It is a perfect embodiment of the architectural adage – so neglected in the hospital construction in Britain of recent years – that the secret of a successful building is an informed client. Dev knew exactly what would make his hospital work efficiently.

There are uniformed guards in military caps at the entrance, who snap to attention as I enter.

'Good morning sir!' they say, whipping off smart salutes. The receptionists, in elegant blue saris, smile at me while pressing their hands together in respectful greeting.

'*Namaste*, Mr Marsh!'

This is rather different from entering my hospital in London in the morning.

Nepal has a very strong caste system. Ritual burning of widows and slavery were abolished only in 1924. Although discrimination on the grounds of caste or ethnicity is illegal, caste is still very important. Nepal was entirely closed to outsiders until the 1950s, and ruled by an absolute, feudal monarchy where the king was believed to be the incarnation of the god Vishnu. The end of the monarchy was precipitated in 2001 by the crown prince taking a submachine gun to his own parents, killing them and several other family members. He was then shot in the head – there are conflicting accounts as to whether he did this himself or not. Dev operated on him, carrying out a decompressive craniectomy, but – I suspect to everybody's relief – he died. There are over a hundred ethnic groups, often with their own languages and castes. It is a nation of immigrants – Mongols from the north and Indians from the south, often living in isolated mountain valleys. It remains a deeply divided and hierarchical society, although most people still look up to foreigners, who are treated with respect, verging on servility. Landlocked, stuck between China and India – described by one of its most famous kings as a 'yam between two rocks' – ethnically so diverse and hierarchical, desperately poor and damaged by the recent earthquake, over-dependent on foreign aid and NGOs, Nepal is a tragic mess. The politics of the country is largely the politics of patronage and corruption, with little sense of the public good and public service which we take for granted in the West. The towns are festooned with advertisements for foreign language courses, promising work abroad. Most Nepalis, if they possibly can, want to leave Nepal. And yet, as an outsider, it is almost impossible not to fall in love with the land and its people.

Can you really fall in love with a country, with a people? I thought that you could only fall in love with a person, but in my first weeks there I started to feel for Nepal as I felt for the women with whom I have fallen in love – seven in total – over the course of my life. Yet I knew that the intensity of my feelings for Nepal would be just as ephemeral as my feelings for the women with whom I had been in love (and much of the love was unrequited anyway). Furthermore, I was leading an utterly spoilt and luxurious life, waited on hand and foot, and in one of the poorest countries on the planet. Some people would probably view my feelings with disdain. But at least I am trying to be helpful and of service, I told myself – not so much with the operating but with trying to help the young doctors become better doctors.

When I was told one morning that the MOs (medical officers) wanted me to stay for ever, I felt very happy and proud. But of course disillusion – or at least a more realistic understanding of Nepal and its sad and intractable problems – was to come quite quickly. There were periods of intense frustration and long periods of inactivity. At times I became deeply despondent. I felt that I was living in self-imposed exile. I often longed to return home, to my family and friends, and wondered why I had abandoned them. I thought of how I had always put work first, ahead of my wife and children, when I was younger, and now I was doing it all over again. But the deep contentment I experienced each day as I walked to the hospital in the low morning sunlight never faded.

I climb up the stairs to the third floor, past the locked suite with the letters VVIP over the door – built in case the president or prime minister falls ill – and go to the library. There are wide windows and on clear mornings you can just see the glittering snow-covered peak of Mount Ganesh, like a broken white tooth above the green hills of the Shivapuri National Park to the north of the city. There is an army base in the park, which was once a TB sanatorium. Some claim that during the recent civil war people were taken there to be tortured, and that many of them disappeared, but others deny it. Nepal has yet to come to terms with its civil war, and the atrocities carried out by both sides. I sit down and wait for the junior doctors to arrive.

The juniors drift in one by one – although the more enthusiastic of the registrars will already be waiting for me. Nepalis are not good time-keepers. About half of the ten medical officers have turned up.

'Good morning everybody,' Salima, the duty MO, says. She is wearing a short white coat and standing in front of a white board on an easel on which there is a handwritten list of the hospital's admissions and discharges. Salima is rather nervous as she knows I am going to quiz her about the cases. She looks a little Chinese, but with enormous black eyes behind a large pair of spectacles. I was to see her at a hospital get-together a few days later, dancing exquisitely to Nepali music. The Nepalese, both men and women, are almost all very good-looking, with a complex mix of Indian, Mongol and Chinese faces. There has been a population explosion in the last thirty years as a result of declining infant mortality, so the streets are full of young people. So many of the men work abroad – 30 per cent of Nepal's national income comes from remittances – that you see far more women than young men on the streets.

'Eighty inpatients, seven admissions, mortality one and no morbidity,' Salima rattles off quickly.

'Well, what's the first case?' I ask.

'Fifty-year-old lady present with loss of consciousness two days ago. Bowels open every day. Known hypertensive and alcoholic. On examination . . .'

'No, no, no! What's she do for a living?' I ask. I have noticed they never describe the patient's occupation, which is supposed to be a normal part of presenting a patient's history, although in Nepal it seems that everybody is either a farmer, a driver, a shopkeeper or a housewife. Mentioning the patient's occupation is important: not so much for the traditional reason, which is to alert us to possible occupational diseases, but more to remind us that the patient is a person, an individual, and has a life and a story beyond being a mere anonymous patient with a disease.

Salima looks embarrassed and fumbles with the sheet of paper in her hand. She probably hadn't seen the patient herself and relied on what had been written by one of the other junior doctors, so I was being unfair.

'Shopkeeper,' she says after a while.

'You're guessing!' I say and everybody laughs, Salima included.

'Now tell us about this loss of consciousness.'

'She comes from other hospital . . .'

'So we have no real history? Whether she had a headache first, whether she fitted?'

Salima looks awkward and says nothing.

Protyush, the registrar who had been on call, takes pity on her.

'Her husband found her on the floor at home. She was intubated at the other hospital and the family wanted her brought here.'

Dev's hospital is a private hospital; patients only come here by choice, or by their family's choice, and only if they can afford it. On the other hand they also have to pay if they go to a government hospital, where the treatment is only free in theory, and possibly worse.

'OK,' I said. 'Salima, what did you find on examination?'

'She localize to pain, not eye opening. Make sounds. Pupils equal and reacting. Cranial nerves intact. Power one on right, plantars up-going,' she continued in high-speed Nepali English, 'CT scan show . . .'

'No, no,' I interrupt again. 'What's your one-line summary?'

'Fifty-year-old lady present with loss of consciousness with known hypertension. Bowels open regularly. On examination pupils equal and reacting, and . . .'

'Salima – one line, not three!'

After a while we agree on a one-line summary. Presenting cases is a hugely important part of medical practice – about both communication and analysis. A short summary after presenting the details of a case forces the doctor to think about the diagnosis. I quickly learnt that most of the doctors were so shy in front of me that they found it very difficult to think analytically. It took them a long time to overcome this in my presence. I also suspected that much of their teaching had been entirely by rote.

'Right, now we can look at the CT scan.'

The scan showed that almost all of the left side of the woman's brain was dark grey, almost black. The woman had clearly suffered a massive and irreversible stroke – an 'infarct' caused by a blood clot forming in the left carotid artery. The left cerebral hemisphere, along with all her language and much of her intellect and personality as well as her ability to move the right side of her body, was dead, with no chance of recovery. Such damage cannot be undone. Some surgeons favour opening up the patient's skull to allow the dead, infarcted brain to swell outwards and stop the patient from dying from the build-up of pressure in the skull, as infarcted brain swells and severe brain swelling kills you.

Helping a patient survive a stroke with this operation of 'decompressive craniectomy' is perhaps justifiable if the stroke is on the right side of the brain (so that they do not lose the ability to communicate, speech usually being on the left side) and if the patient is young, but it seems a strange thing to do in patients who are going to be left dreadfully disabled if they survive. And yet it is recommended in articles in various learned journals claiming that such patients are happy to be alive, and is widely practised. You might wonder how the victims' happiness can be established if they have lost much of their intellect and personality, the part of their brain responsible for self-respect, or the ability to speak. You might also wonder whether their families are of the same opinion as the patients. Patients with severe brain damage, as far as you can tell, will often have little insight or understanding of their plight, whereas those that do are often deeply depressed. In a way, the true victims are the families. They must either devote themselves to caring, twenty-four hours a day, for somebody who is no longer the person that they once were, or suffer the guilt of consigning them to institutional care. Many marriages fail when faced with problems of this sort. It is worst for parents, who are tragically bound to their brain-damaged children, whatever their age, by unconditional love.

'So the patient's going to die?' I ask the room at large.

'We operated,' Protyush says. I express surprise.

'I spent half an hour trying to persuade the family that we shouldn't operate but they wouldn't accept it,' he adds.

After the morning meeting I go downstairs, take my shoes off outside the operating theatre and ITU area, get the uniformed guard to open the locked door for me and choose a pair of ill-fitting pink rubber clogs from a rack in the theatre corridor. Nepali feet are mostly small so I hobble uncomfortably to Dev's office, which is conveniently located between the ITU and the theatres.

Dev and I had always got on well together as colleagues when we were training together thirty years earlier, but it had been little more than that. I regret to say that I was far too ambitious and concerned for my own career at that time to take much interest in my colleagues, although I suppose that working a 120-hour week and having three young children at home left me with little spare time. And yet as soon as I came to Kathmandu, Dev and his wife Madhu were so welcoming that it felt as though we had always been the oldest of friends, even though we had only seen each other briefly at a few conferences over the intervening years. Dev is also charismatic, a man of great integrity and very determined. Like most Nepalis he is quite short and slight, although now a little rounder (which he blames on my presence and the beer we drink in the evening). He has a prominent, stubborn chin but slightly hunched shoulders, so that he looks like a cross between a bulldog and a bird. His intensely black, wavy hair has now turned grey. He has a chronic cough which he attributes to breathing the polluted air of the city centre when he worked for many years in the government hospital known as the Bir. He speaks very fast, with great animation, as though in a permanent state of excitement, about his past achievements and the great difficulties he had to overcome in trying to bring neurosurgery to Nepal. He also talks of how difficult it is to run a major neurosurgical practice more or less single-handedly.

He told me that it had been much easier when he was the only neurosurgeon in Nepal – if he gave bad news to patients they had little choice but to accept it. But now there are other neurosurgeons, most of whom have worked with him, from whom they get second opinions, and it would seem that there is little love lost between the professor and some of his former trainees. So he sometimes now has major problems with patients' families when things have gone badly, which they so often do with neurosurgery. When he told me this, I pointed out that in England there was more and more litigation against doctors, and this always involved doctors giving evidence against each other as expert witnesses.

'Yes, but here the families threaten us with violence, demand money and have even said that they'll burn the hospital down,' he retorted. 'Of course, we don't really have malpractice litigation here – it's almost unheard of to sue doctors.'

Doctors, especially surgeons, are often intensely competitive, and we all worry that other doctors might be better than we are, although I can think of a few famous international surgeons who are so supremely arrogant that they seem to have suppressed this problem by completely forgetting their bad results. We need, of course, self-confidence to cope with the fact that surgery is dangerous and we sometimes fail. We also need to radiate confidence to our frightened patients, but deep down most of us know that we might not be as good as we make out. So we feel easily threatened by our colleagues and often disparage them, accusing them of having the faults that we fear we have ourselves. It is made all the worse if we surround ourselves with junior colleagues whose careers depend on us, and only tell us what they think we want to hear. But it is also because, as the French surgeon René Leriche observed, we all carry cemeteries within ourselves. They are filled with the headstones of all the patients who have come to harm at our hands. We all have guilty secrets, and silence them with self-deception and exaggerated self-belief.

Dev remembers all sorts of details of our time spent working together in London, which I have long forgotten. His determination and energy are remarkable and I quickly came to understand why he has had such an extraordinary and brilliant career and is famous throughout Nepal. This has not been without its disadvantages. Driven and ambitious people can achieve great things, but often make many enemies in the process. Patients come to his outpatient clinic with all manner of non-neurosurgical problems, hoping that he can cure everything. A few years ago one of his daughters was abducted from the family home at gunpoint and Dev had to pay a large ransom. Since then he goes everywhere with a bodyguard.

The ITU is a large room with good natural light, as there are windows all the way along two of the walls. There are ten beds; they are rarely empty. The hospital admits strokes as well as head injuries and many of these patients have undergone decompressive craniectomies. Most of the patients are on ventilators, with pink bandages around their heads and the usual array of monitors and drip-stands and flashing lights and noisy alarms beside them. I had forgotten how grim neurosurgical ITUs can be – in London I had only been responsible for a small proportion of the patients since I was only one of many consultants.

Many of the patients on the ITU here would not survive, few would make a good recovery, especially in Nepal.

'You do far more decompressive craniectomies here than I would do,' I say to Dev. 'Only in America have I seen so much treatment devoted to so many people with such little chance of making a useful recovery. And yet Nepal is one of the poorest countries in the world.'

'I have to compete with many other neurosurgeons – trained in India or China – and they'll operate on anything, and it's always for the money. Like in America. If I tell the family now that no treatment is possible, they'll go and see somebody else who'll tell them the opposite and then they'll kick up a big fuss. So I am forced to operate now when in the past I wouldn't have. I often wish I still worked for the NHS,' he adds.

My colleague Igor in Ukraine often faced similar problems. I have been in countries where the surgeons sometimes have to operate with the patients' families outside the operating theatre wielding guns, threatening to kill the surgeon if the operation is unsuccessful. As a visiting doctor from the West it is hard, at first, to understand the difficulties our colleagues face working in countries with very different cultures and without the rule of law. It is easy to feel superior, to pass condescending judgement. I hope that over the years I have learnt to observe, and no longer to judge. I want to be useful, not to criticize. Besides, so often I find that I have misunderstood or misinterpreted what I have seen or been told – I have learnt not to trust myself. All knowledge is provisional.

'Many of these patients are going to die anyway, aren't they?' I say as we look at the next comatose patient with a bandaged head, labelled 'No Bone Flap'. After a decompressive craniectomy the patients are left, for a few weeks or months, with a large hole in their skull, like a giant version of the fontanelle with which we are all born. The 'No Bone Flap' label is to remind the medical and nursing staff that part of the brain is no longer protected by overlying bone. This particular patient – like so many in Nepal – has been involved in a motorbike accident.

'Cultural case,' Dev says. 'The family ties here are so strong. The family just can't accept that there is no treatment. If I hadn't got the boys to operate last night the family would say: "Oh Neuro Hospital doesn't want to operate!" Can you imagine the situation? Next thing they take the patient out of my hospital and somebody else will operate. The patient will be a vegetable but the family are happy and my reputation will be rubbished . . .'

Dev turns to look at me.

'When I was the Minister of Health under the last king – before the Maoists abolished the monarchy – I saved more lives by making crash helmets for motorcyclists compulsory than I will ever save as a neurosurgeon. Most of the families are uneducated,' he goes on. 'They have no conception of brain damage. They are hopelessly unrealistic. They think that if the patient is alive they might recover, even if the patient is just about brain-dead. And even if they are brain-dead they still won't accept it.' I was to learn more about this later.

So much for the value of commercial competition in health care, I thought, in a poor country like Nepal. And all this on one man's shoulders, day in, day out, with never a day off, for thirty years.

Neurosurgery is something of a luxury for poor countries. Illnesses requiring neurosurgical treatment are relatively rare compared to problems affecting other parts of the body. It requires very expensive equipment, and for problems such as cancer and severe head injuries treatment often fails or achieves little. We operate in the hope that patients will make a good recovery, and many will. There can be wonderful triumphs, but the triumphs wouldn't be triumphant if there weren't disasters. If the operations never went wrong, there would be nothing very special about them. Some patients will be left more disabled than they were before surgery and others, who would have died if we had not operated, will survive, but terribly disabled. At times, in my more despondent moments, it is not always clear to me whether we are reducing the sum total of human suffering or adding to it. So for countries like Nepal and Ukraine, with impoverished and weak governments and poor primary health care, it makes little sense to spend large sums of money on neurosurgery. Dev in Nepal and Igor in Ukraine have had little choice other than to move into private practice, albeit reluctantly, and yet both feel a little tainted by it, even though they often treat poor patients for free. But there is a limit to how often you can do that if your hospital is to survive.

There has always been a tension at the heart of medicine, between caring for patients and making money. It involves, of course, a bit of both, but it's a delicate balance and very easily upset. High pay and high professional standards are essential if this balance is to be maintained. The rule of law, after all, in part depends on paying judges so well that they will not be tempted to accept bribes.

Many medical decisions – whether to treat, how much to investigate – are not clear-cut. We deal in probabilities, not certainties. Patients are not consumers who, by definition, always know what is best for themselves, and instead must usually accept their doctors' advice. Clinical decision-making is easily distorted by the possibility of financial gain for the doctor or hospital, without necessarily being venal (although it certainly can be). Increasing litigation against doctors also drives over-investigation and over-treatment – so-called 'defensive medicine'. It is always easier to do every possible test and treat 'just in case' rather than run the risk of missing some very obscure and unlikely problem and being sued. This combination of paying doctors on a 'fee for service' basis – the more we do, the more we get paid – and increasing litigation against doctors in many countries is one of the reasons why health-care costs are running out of control.

On the other hand, a fixed salary can breed complacency and an irritating moral righteousness, to be found in some doctors who disdain private work. It is indeed a delicate balance, and Dev and Igor, both doctors of great integrity, have mixed feelings about running private hospitals.

'I am the country's highest taxpayer,' Dev tells me with a laugh, pointing to a photograph on his office wall of the Finance Minister recently handing him a certificate to this effect. Yet it seems highly unlikely to me that Dev is the highest earner in Nepal.

In Nepal and Ukraine – and many other countries – government is widely seen as corrupt and, understandably, people are reluctant to pay taxes, doing everything they can to evade them. There's another parallel here between Dev and Igor: both are scrupulous in paying their taxes. But it is difficult to be honest in a dishonest society, and many people will hate you for it.

Low tax revenues mean that governments in poor countries like Nepal and Ukraine have little money to spend on health care and infrastructure projects that would benefit the country. Besides, Ukraine is involved in a war and Nepal is still recovering from a vicious civil war. The lack of government spending on welfare and infrastructure only serves to reinforce the public's reluctance to pay taxes. It is a vicious circle from which it is very hard to escape. Driving in

Kathmandu can be a vision of hell and Hobbesian anarchy, especially at night in the suburbs. There is no street lighting. Trucks, cars and motorbikes are crammed together in narrow, rough lanes, driving in a cloud of dust and diesel fumes, eerily lit by undipped, dazzling headlights. Nobody gives way, each driver tries to go first – if you give way you will never move. There is no argument or shouting, nobody loses their temper, there is only the occasional blowing of horns. Everybody is resigned to the grotesque struggle which they have no power to end. Pedestrians join the crush to cross the road like ghosts in the dust. The unfortunate traffic police must inhale the poisoned air all day when they stand at the crossroads, trying to direct the chaotic vehicles. The city is asphyxiating, but the government appears to be utterly helpless and apparently has no plans to do anything about it at all.

The only certainties in life, as Benjamin Franklin once observed, are death and taxation. We all try to avoid both. But health care is getting more and more expensive – in most countries the population is ageing and needs more medical attention, and high-tech modern medicine is ever more extravagant. We all want to see cancer cured, but this will only drive costs up and not down. Not just because the complex genetic and drug treatments involved are so costly but because more of us will then live longer, to die later from some other disease, or slowly from dementia, requiring constant and expensive care. And rather than discover new antibiotics – the human race, especially in poor countries, faces decimation within a few decades from bacterial antibiotic resistance – the pharmaceutical companies concentrate on drugs for cancer and the diseases, such as diabetes and obesity, of affluence.

So health care is becoming ever more expensive, but most

governments fear that putting up taxes or insurance premiums will lose them the next election. So instead, in the West, a small fortune is spent on management consultants who subscribe to the ideology that marketization, computers and the profit motive will somehow solve the problem. The talk is all of greater efficiency, reconfiguring, downsizing, outsourcing and better management. It is a game of musical chairs where, in England at least, the music is constantly being changed but not the number of chairs, and yet there are more and more of us running around the chairs. The politicians seem unable to admit to the public that the health-care system is running out of money. I fear that the National Health Service in England, a triumph of decency and social justice, will be destroyed by this dishonesty. The wealthy will grab the chairs, and the poor will have to doss out on the floor.

As the weeks went by I took to absenting myself from the ITU rounds, unless there was a patient with whose operation I had been involved. I found the rounds too depressing.

After the ITU round Dev spends up to an hour on 'counselling'. The patients' families will stay in or near the hospital throughout the time their family member is there. There is a small hall in the centre of the hospital's first floor, well lit by a glass roof and decorated with palms in large planters. A prayer room with colourful Hindu and Buddhist icons is on one side. The families of the patients on the ITU wait here to be seen, one by one, by Dev and his colleagues in the counselling room next to the prayer room. They are updated on their relative's condition, questions are answered, and then they sign the medical notes, confirming what they have been told.

'I had problems to begin with,' Dev said. 'Some of the

families denied that they had had things explained to them, so I now do it formally every day.'

Although it was all in Nepali, it was fascinating to see Dev at work. As all good doctors do, he adjusted his style to the people he was talking to – sometimes joking, sometimes grave, sometimes consoling, sometimes dictatorial. On one occasion the patient's daughter was a nurse who had been working in England and spoke good English. Her elderly mother had suffered a huge stroke and the whole of the right side of her brain had died. She had undergone decompressive surgery and had therefore not died within the first few days but was now lying in the ITU, half paralysed and unconscious.

'You talk to her,' Dev muttered to me, 'and you'll see the problem.'

So I spoke to the daughter as I would speak to the families of my patients in England. I told her that if her mother survived she would be utterly dependent and disabled, with grave damage to her personality and intellect.

'Would she want to survive like that?' I asked. 'That's the question you and the rest of the family should be asking yourselves. I would not want to live like that,' I added.

'I hear what you are saying,' she replied, 'but we want you to do everything possible.'

'You see?' Dev said to me later. 'They're all like that. I've even had it with families of doctors. They just can't face reality.'

The child's head was completely shaven and had already been fixed to the operating table with the pin headrest. The juniors had had problems inserting a central intravenous line into one of the major veins in the neck and had ended up hitting the carotid artery. They then decided to rely on two large peripheral lines in the smaller veins of her arms for blood transfusion, in case there was heavy bleeding from the tumour. So there had been long delays before I came into the operating theatre. Much of her face was hidden by the plaster strapping holding the endotracheal tube in place, but despite all this and the disfiguring shave, she looked painfully sweet and vulnerable, with a broad Tibetan face, light-brown skin and slightly red-tinted cheeks.

Dev was standing by the patient's head. 'You and I trained together,' he said. 'We think along the same lines.' He has six trainees whom he has trained to do the simple emergency work and the 'opening and closing' of the routine surgery. Dev, however, does almost all the major operating himself. Occasionally he has been joined by foreign surgeons, but only for short periods of time. There are major cases to be done every day, six days a week, and the pressure is relentless. In six weeks working in Kathmandu I saw more major operations than I would have done in six months in London.

This was the first time I had seen the child, although I had looked carefully at her brain scan with Dev earlier that morning.

'She was operated on by one of the other neurosurgeons here in Kathmandu,' Dev told me, 'but I don't think he removed much of it. Just did a biopsy. It's said to be a Grade Two astrocytoma.'

'It's not a good tumour,' I said, looking unhappily at the scan. 'It may be benign but it's involving all the structures around the third ventricle and God knows where the fornices are.'

'I know,' said Dev.

The fornices are two narrow bands of white matter, a few millimetres in size, that are crucial for memory. White matter consists of the billions of insulated fibres – essentially electrical cables – that connect the eighty or so billion nerve cells of the human brain together. If the fornices are damaged, people lose a large part of their ability to take in new information – a catastrophic disability.

Average income in Britain is forty times greater than in Nepal. Primary health care in Nepal is poor (although better than in many other low-income countries) and diagnosis of rare problems such as brain tumours is invariably delayed. The tumours, therefore, by the time they are diagnosed, are much larger than in the West and treatment is more difficult, more dangerous and less likely to achieve a useful result. Brain tumours in children are very rare but very emotive, and although the rational part of myself considered that operating on this child was a waste of time and money, it is almost always impossible, wherever you are in the world, to say this to the desperate parents. And I myself had once been the parent of a child with a brain tumour. But the decision was Dev's responsibility and not mine.

Once I had checked that they had positioned the child correctly, I left them to start the operation, returning when a nurse came to Dev's office and silently beckoned me to come to the operating theatre and join Dev.

I am becoming little better than a vet, I told myself as I scrubbed up at the long zinc sink with its row of taps and iodine dispensers. I am operating on patients without knowing anything about them, without even seeing them other than as unconscious, impersonal heads in a pin headrest.

4

AMERICA

One year before I went to Nepal, and before I had retired, I attended a cerebrovascular workshop in Houston, intended to help trainee surgeons learn how to operate on the brain's blood vessels. I was to be one of the instructors. I arrived from London after a ten-hour flight. The workshop started next morning at eight, after I had delivered a lecture at seven to my colleagues in the neurosurgical department which I was visiting. American hospitals start early – the interns, the most junior doctors, often begin their ward rounds before five in the morning. I once asked a group of them about the physiological effects of sleep deprivation on their patients and they seemed quite startled by the suggestion that their immensely hard work might actually be harming the patients.

My lecture was about how to avoid mistakes in neurosurgery, but only a handful of people had turned up to listen to me, presumably because they felt that they had little to learn from the mistakes made by an obscure English surgeon such as myself. The large breakfast laid out in the room outside the lecture theatre remained uneaten. There was a short briefing at the beginning of the workshop. We sat on tiered seats in a small room with three enormous LED screens in front of us. Everything looked new and immaculately clean. A businesslike woman dressed in scrubs told us that under no circumstances was photography permitted and that everything we would be doing was regulated by federal law. She cited various specific statutes which were flashed up on the screens, each one with a long reference number. She also told us that we must respect the subjects of the workshop. Different-coloured hats were then handed out – mine was blue as I was a member of the faculty. The medical students' were yellow and the neurosurgical residents' were green. We were then ushered through a pair of large security doors into the research facility.

This looked like a cross between an operating theatre and an open-plan office, with several bays. Floor-to-ceiling windows looked out onto the many glittering skyscrapers that form the Texas Medical Center, the largest concentration of hospitals anywhere on the planet. There are 8,000 hospital beds here – fifty-one clinical institutions in total, I was told – practising some of the most advanced medical care anywhere in the world. There were half a dozen shapes lying on operating tables; I suppose each one was about the size of a ten-year-old child. They lay entirely hidden under blue surgical drapes, with anaesthetic tubing and cables coming out of one end, connected to the same ventilators and monitors with colourful digital displays that I see every day at work. I walked up to one of them and put a hesitant hand out – it was strange to feel the hoofed trotters under the drapes at the end of the operating table.

'Isn't this just fantastic!' said my colleague, a trainee of mine from many years ago, who had recently become the chairman of the neurosurgical department which was staging the workshop, which he had organized. 'Nobody anywhere else is doing anything like this! Come on, guys!' he said to the residents in their green hats. 'Enjoy!'

One of the faculty pulled back the blue drapes off the head of one of the pigs and started to operate. The pig was lying on its back with its broad, pink neck stretched out. It had probably been shaved, and although it was clearly not a human neck – it was far too flat and wide – the skin looked disconcertingly similar. He used cutting diathermy to dissect down to the carotid artery, one of the main arteries for the brain – a smaller vessel than in a human. The plan was to dissect out a vein and graft it to the artery, creating an aneurysm, a model for the life-threatening aneurysms that occur in people and cause fatal haemorrhages. The artificial aneurysm can then be treated – with an 'endovascular' or 'coiling' technique where a microscopic wire is inserted into the aneurysm via the artery, involving only a simple puncture in the skin, and the aneurysm is blocked off from the inside. Alternatively the aneurysm can be treated with the more old-fashioned method of open surgery, where it is clipped off from the outside. Most aneurysms in people in the modern era are treated with coils, but a few still need clipping. The purpose of the workshop was to give trainee surgeons some practice in the techniques without putting a human life at risk. I am sentimental about animals, and felt sorry for the pigs, but reminded myself that they were doing more for humanity by being used for surgical practice than by being turned into bacon – and there were all those federal statutes protecting them, after all.

My fellow instructor started to stitch the vein graft to the artery. It was a rather slow business and I wandered off towards a group of doctors gathered in a corner of the room. A blue-capped faculty member was talking with great enthusiasm.

'This is awesome! This is so much better than specimens preserved in formaldehyde!'

I looked over his shoulder. Two trainees were operating on a severed human head. It was held in the steel head clamp most neurosurgeons use when operating and the skin of the neck had been formed into two flaps; these had been stitched together with a few broad sutures to form a stump, although some slightly obscure fluid was dripping out between the sutures. If I had not done my year of cadaveric dissection as a medical student forty years ago I think the sight would have given me nightmares for many days afterwards. It was bizarre and disturbing to see a head in a standard head clamp – something I must have experienced thousands of times with living patients when I operate – and yet with no body attached to it.

So I joined the small group standing around the two trainees who were carrying out a craniotomy under the guidance of a fellow instructor – sawing open the severed head with surgical tools, looking down an expensive operating microscope. I was staggered by all the equipment which surrounded the various stations, six with anaesthetized pigs and now one with a dead person's head. All of it had been provided by the manufacturers – hundreds of thousands of dollars' worth, all to be used for practice. As I watched the two trainees uncertainly drilling into the severed head, a young man behind me – dressed, to my surprise, entirely in black scrubs like a ninja – accosted me.

'Professor!' he said, with the passionate conviction of an equipment rep. 'Have a look at this.' He pointed to the beautiful array of miniature titanium plates and screws and tools, each in its own perfect moulded cavity on a black plastic tray in front of him. These plates are screwed in place to reassemble the skull after sawing it open – although in this case only for practice.

'Have you tried our latest electric screwdrivers?' he asked, handing me a neat little battery-powered screwdriver which I suppose would save about five seconds, and needed only marginally less effort than the manual screwdriver I normally use when putting patients' skulls back together again with titanium plates. I switched the electric screwdriver off and on, marvelling at the extravagance of the American medical system.

'How d'ya like it?' asked the rep.

'Outstanding,' I replied, thinking of how, on my flight the previous day, the pilot had told us over the intercom when the plane was about to begin its descent that now would be an outstanding time to visit the restrooms.

'Guys! We have a master here!' the instructor called out when he saw me. 'Professor, can you give us some surgical pearls?' I thought a little apologetically of the swine in the nearby bay undergoing surgery.

Happy to have something useful to do, I pulled on a pair of gloves and went up to the microscope to reposition it and look down into the dead brain.

'Have you got any brain retractors?' I asked. '"Ribbons" you call them here in the US.' It seemed they did not, so I used a small chisel to gently lift up the frontal lobe. There was, of course, no bleeding, but the consistency of the dead tissue was not unlike that of the living thing.

'Formaldehyde makes it all stiff and solid, and it smells awful,' I said. 'But where do they get these freshly dead heads?' I asked of nobody in particular.

'Maybe a John Doe scraped off the sidewalk,' somebody offered.

Using the small chisel I dissected out the anterior cerebral arteries, explaining how you approach an anterior communicating artery aneurysm by resecting – that is, removing – part of the brain called the gyrus rectus to find the aneurysm.

'The gyrus rectus serves olfaction,' I told my small audience. 'The patients are better off with perhaps some impairment of smell than dying from another haemorrhage if they don't have the aneurysm treated.'

I handed over the operation to the two residents and walked round to look at the dead face: head shaved, eyes closed, stubble on his cheeks, blackened stumps of a few remaining teeth. He clearly had never seen a dentist. As far as I could tell he was not – or rather had not been – that old before he died. It was impossible not to wonder for a moment who he had been and what sort of life he had led, and to think that once he had been a child, with all his future in front of him.

Workshops like this are not unusual, but I had never been at one before and I found it rather distressing. I would consider this to be a weakness on my part – it is clearly much better that trainee surgeons should practise in workshops like this than on living patients. When I was back in England two weeks later I mentioned this to a colleague who had recently organized a similar workshop in the UK.

'Ah!' he said with a laugh. 'Only one? I had fifteen heads, freeze-dried, flown in from the US for my skull base workshop last year. I needed to put them all through the MRI scanner before the meeting and drove to the hospital with the heads in the boot of my car. I wasn't quite sure what I would say if I was stopped by the police. The other problem was that they were starting to thaw. I don't know where they get them from,' he added.

I left the room with its severed head and anaesthetized pigs and found another huge breakfast laid out next to the lecture theatre where we had started. After breakfast I was taken on a whirlwind tour of the hospital.

The hospital consisted of a series of multi-storey towers, and we went through what seemed to be an almost endless series of huge lobbies and halls. The hospital had its own twelve-floor hotel; patients came from all over the world for treatment, not just from America. There were twenty – twenty! – other hospitals next to my colleague's, as well as many other medical and clinical research institutions. The Medical Center occupies more than a square mile, and when I looked out of my twelfth-floor hotel window all I could see was hospital after hospital, all built of glittering glass, receding into the distance like a mountain range. Medicine in the USA is notoriously extravagant. I saw one hospital in Chicago which had a luxurious restaurant, bar and garden on its roof. The hospitals are locked in fierce competition for business and many are designed to look as little like hospitals as possible. They resemble instead luxury hotels or shopping malls or first-class airport lounges. They are the peacocks' tails of health care.

That evening my colleague took me to his country club. We drove there through the city's suburbs, past large mansions with pillared porticoes and extensive lawns. The club too was built on a grand scale and the clubhouse – icy-cold with air-conditioning – had a massive baronial fireplace in the Scottish style decked out with mounted stags' heads on either side, and a large reproduction of the famous Victorian painting by Landseer of a stag, known as *The Monarch of the Glen*, hung above the grand staircase. We had an excellent dinner there. The waiters were elderly Mexican men with solemn and expressionless Aztec faces. They were dressed in black suits with white aprons and they moved with slow dignity as they served the clientele, nearly all of whom were dressed in baggy shorts and long T-shirts. Over dinner there was the usual surgical gossip – mainly about a colleague who had been sacked for having an affair with a rep, and whether the rep was enhanced with silicone or not. Opinions differed as to this latter question. After his dismissal she had apparently sued him successfully for sexual harassment but now, my colleague told me, they were back together again. I also learnt that the operation on the pig to create an aneurysm had not been a success: one of the technicians had forgotten to give the animal an anti-coagulant injection and the pig had suffered a major stroke as a result of the surgery to its carotid artery. It would, however, have presumably been sacrificed – as it is called – in any case, even if the mistake had not occurred.

After dinner we went out into the sweltering, humid heat to inspect a car show outside the clubhouse. Thirty or so classic cars were drawn up in the car park, shiny and polished, many with their bonnets up so that you could see the spotless, chromed engines inside. A red Ferrari inched its way past us to find a parking place.

My colleague nudged me and said with awe: 'That's a seven-million-dollar car. And the guy driving it is a billionaire.'

It transpired later that the car was only a reproduction, but was still worth a million dollars. The billionaire apparently was a real billionaire but looked a fairly ordinary sort of guy. A group of people gathered admiringly around the car once the billionaire had parked it, and they took photographs of each other in front of it.

I went out for a run next morning as the sun was rising. I was streaming with sweat within a matter of minutes as I ran along the street beneath the tall hospital towers, past neatly tended flowerbeds. At the edge of the great block of hospitals there was a large park, with a miniature railway line running round it. Several dozen homeless people were dossing out on the benches and sidewalks in one corner of the park. I was told later that there was a church nearby which gave out free meals. As I ran back to the hotel the sun rose behind me, over the dozens of buildings of the Medical Center, and I was almost blinded by its dazzling reflection in the thousands of hospital windows facing me.

5

AWAKE CRANIOTOMY

For a surgeon to help operate on patients he did not know, whom he would scarcely ever see again, for whom he carried no practical responsibility – if there were problems Dev would have to deal with them – had always been anathema to me. And yet I had already discovered, to my surprise, that my lack of human contact with the Nepali patients both before and after surgery had not reduced my anxiety when I was operating. It did not seem to matter after all. Operating in Kathmandu I was in the same state of tense concentration as I was in London and it seemed that I cared just as much for the patients, even though my concern for them had now become entirely abstract and impersonal. I used to feel critical of surgeons if they were remote and detached from their patients but now, very late in my career, I was forced to recognize that some of this had perhaps been vanity on my part, and simply yet another attempt to feel superior to other surgeons.

Surgeons describe operating on patients with whom they have no personal or emotional contact as being veterinary. There was a veterinary surgery near the old hospital in Wimbledon and one of the vets there – Clare Rusbridge

– specialized in veterinary neurological disorders. Devoted pet owners can take out insurance for their pets which includes the cost of MRI brain and spinal scans. Clare would bring to our weekly X-ray meetings fascinating scans of cats and dogs with neurological disorders. We would look at the scans at the end of the meeting and called it Pets' Corner. They provided a bizarre contrast in anatomy to the images of human brains and spines with which we were so familiar. Cavalier King Charles spaniels, we learnt, often suffer from the brain abnormality known as a Chiari malformation, which humans also get. Labradors can develop malignant meningiomas. The spaniels' problem is the result of selective breeding aimed to produce the small round head which wins points at dog shows. The malformation leads to spinal cord damage, and the poor creatures suffer from intractable pain and scratch themselves incessantly.

I operated with Clare on a couple of occasions, though she was unable to find an owner of a King Charles spaniel who was willing to let us operate on their pet. We did, however, once operate on a badger, which had been found confused and wandering on Epsom Downs and had been rescued and brain-scanned by an animal charity. The brain scan suggested that she might have hydrocephalus, although, to be honest, not much is known about badger brains. She was a beautiful creature and once she had been anaesthetized, I held her on my lap for a few minutes, stroking her grey and white fur, before Clare removed most of it with a pair of clippers in preparation for the surgery. I tried to carry out an operation for the possible hydrocephalus. I already had an article published entitled 'Brain Surgery in Ukraine' and I hoped I would be able to add to my CV 'Brain Surgery in Badgers', but the operation was not a success and the poor creature died. Or rather, she was 'euthanased'.

'At least our patients don't have to suffer,' one of Clare's colleagues, who had watched the operation, commented afterwards. 'Unlike yours.'

The first case I had done with Dev – two days before the operation on the child – had been an awake craniotomy for a tumour. This was the first time that such surgery had been carried out in Nepal. I had brought the equipment for cortical brain stimulation from London in my suitcase. Many years ago I had been the first surgeon in Britain to use the technique of awake craniotomy for treating a particular type of brain tumour known as a low-grade glioma. It was unorthodox at the time, but is now standard practice in most neurosurgical departments. It is, in fact, a very simple way of operating which allows you to remove safely more of a tumour in the brain than with the patient asleep under a general anaesthetic. The problem is that the 'tumour' is in fact part of the brain which has tumour growing in it – brain and tumour are muddled up together. The abnormal area, especially at its edges, looks almost identical to normal brain and only by having the patient awake, so that you can see what is happening to them as you remove the tumour, can you tell if you are straying into normal brain and running the risk of causing serious damage. Patients tolerate the procedure much better than you might expect, once they understand why it has been recommended.

The brain cannot feel pain: pain is a sensation created within the brain in response to electrochemical signals sent to it from the nerve endings in the body. When I see patients with chronic pain, I try to explain to them that all pain 'is in the mind' – that if I pinch my little finger, it is an illusion that the pain is in my finger. It is not 'in' my finger but really in my brain – an electrochemical pattern in my brain, in a map that my brain has made of my body. I try to explain this in the hope that the patient will understand that a psychological approach to pain might be just as effective as a 'physical' treatment. Thought and feeling, and pain, are all physical processes going on within our brains. There is no reason why pain caused by injury to the body to which the brain is connected should be any more painful, or any more 'real', than pain generated by the brain itself without an external stimulus from the body. The phantom limb pain of an amputated arm or leg can be excruciating. But most patients with chronic pain problems or conditions like chronic fatigue syndrome find this hard to accept. They feel that their symptoms are being dismissed – as they often are – if it is suggested that there is a psychological component to their problem and that a psychological approach might help. The dualism of seeing mind and matter as separate entities is deeply ingrained in us, as is the belief in an immaterial soul which will somehow outlive our bodies and brains. My 'I', my conscious self, writing these words, does not feel like electrochemistry, but that is what it is.

So, for an awake craniotomy, only the scalp needs to be anaesthetized and the rest of the operation is painless, although patients find having their skull drilled into very noisy – the skull acts like a sounding board. I therefore usually do this part of the operation under a brief general anaesthetic. The patient is then woken up, but unlike normal operations, where you wake up in a bed back on a hospital ward, with an awake craniotomy you wake up in the operating theatre, in the middle of the operation. There are various ways of conducting the 'awake' part of the operation. All involve using an electrode to stimulate the patient's brain, which tells you where, in functional terms, you are on its surface. You will be able to produce limb movement or interfere with the ability to talk as the electrode momentarily stimulates or stuns the relevant part of the brain. It is a little like pulling the strings of a puppet. You also need to ask the patient to perform simple tasks or name and identify pictures if the tumour is near the speech centres of the brain. Some surgeons rely on speech therapists or physiotherapists to talk to the patient and assess them as the operation proceeds. I always relied on my anaesthetists, in particular Judith Dinsmore, whose highly skilled and reassuring manner never failed to keep the patients calm and cooperative.

I operate with a transparent screen between myself and the patient. Judith would sit facing the patient, talking to them and assessing the relevant functions – their ability to talk fluently, or to read, or to move the limbs on the opposite side of their body to the tumour (for obscure evolutionary reasons, each half of the brain controls the opposite side of the body). I would be standing behind and above the patient's opened head and exposed brain, and watch and listen to Judith through the transparent screen as she put the patient through their paces. When she started to look anxious, I knew it was time to stop. If the patient had been under a general anaesthetic for all of the operation, I would have had to stop much earlier and would have removed less of the tumour. There would have been no way of knowing whether I was still removing tumour or normal, functioning brain. Obviously, more subtle social or intellectual functions cannot be tested, but this is not usually a problem. It would seem that low-grade gliomas have to be very extensive indeed – and effectively inoperable – before the patient's personality is at risk.

I operate with a microscope which has a camera connected to a video monitor. The operation is mainly done with a simple sucker or an ultrasonic aspirator (which is a sucker with an ultrasonic tip that emulsifies what you are operating on). All you can see, as you look into the patient's brain with the microscope, is the brain's white matter, which is like a smooth, thick jelly. It is usually – but not always – slightly darker than normal because of the presence of tumour within it. It took me some time to learn to operate like this, with the patient awake. I am always a little anxious when operating and, at first, having the patient awake made this worse, especially as I had to affect a complete calm and confidence for their sake that I did not inwardly feel.

'Do you want to see your brain?' I will usually ask the patient. Some say yes and some say no. If they say yes, I go on to say: 'You are now one of the few people in the history of the human race who have seen their own brain!' And the patients gaze in awe at their brains on the monitor. I have even had the left visual cortex – the part of the brain responsible for seeing things on the right-hand side – looking at itself. You feel there should be some philosophical equivalent of acoustic feedback when this happens, a metaphysical explosion, but there is nothing, although one patient, having looked at his speech cortex, as I brushed it with a sucker and told him that was what was talking to me, commented: 'It's crazy.'

Towards the end of the first ever awake craniotomy in Nepal, the patient's leg had suddenly become paralysed.

'It's probably temporary,' I assured Dev. 'It can happen when operating in the supplementary motor area, which was where the tumour was.'

I nevertheless awoke next morning feeling miserable. But Dev came to find me as I sat with a cup of coffee in the garden of his home – I had only stayed two nights in the hotel to which I had first been taken, and was now living in the guest house at the end of Dev's garden. He told me that the juniors had rung to say that the patient had started to move his leg.

'I knew you were upset, though you said nothing,' he said.

The morning was instantly transformed.

'Were there any admissions overnight?' I asked.

'Couple of head injuries,' Dev replied.

I would often be rung at night when I was on call in London, although unlike Dev I was not on call every night. The telephone would ring and I would be dragged out of sleep, often with the strange illusion that I had chosen to wake up before the phone started ringing. These emergency cases were usually cerebral haemorrhages – bleeding into the brain caused by head injuries or a weakened blood vessel. I had to decide whether the patient should be operated on or not. Sometimes it was obvious that they would die if they did not undergo surgery and that they would make a good recovery after surgery. Sometimes it was obvious that they did not need surgery and would survive without it, and sometimes it was obvious that they would die whatever we did. But often it was not clear whether to operate and, if you did, whether they would make a good recovery. If the haemorrhage had been a big one, the patient was going to be left disabled, however well the operation went, as the brain – being so intricate and delicate – has much less capacity for repair and recovery than other parts of the body. The question then was whether the disability might be so severe – the patient left a 'vegetable', as the saying goes – that it might be kinder to let them die.

You can rarely predict with absolute certainty from a brain scan what sort of recovery the patient might make, but if we operate on everybody (as some surgeons do), without any regard to the probable outcome, we will create terrible suffering for some of the patients, and even more for their families. It is estimated that there are 7,000 people in the UK in a 'persistent vegetative or minimally conscious state'. They are hidden from view in long-term institutions or cared for at home, twenty-four hours a day, by their families. There is a great underworld of suffering away from which most of us turn our faces. It is so much easier to operate on every patient and not think about the possible consequences. Does one good result justify all the suffering caused by many bad results? And who am I to decide the difference between a good result and a bad result? We are told that we must not act like gods, but sometimes we must, if we believe that the doctor's role is to reduce suffering and not just to save life at any cost.

'Twenty-six-year-old. Collapsed last night while in the shower. Looks like a spontaneous ICH. Probably an underlying AVM – there's some calcification. Blown a big hole in the left basal ganglia, a bit into the midbrain too. GCS four according to the paramedics. Blown left pupil but came down with mannitol and ventilation. Lots of shift on the CT. Basal cisterns just visible. Now tubed and ventilated.'

'Hang on a mo, I'll have a look at the scan,' I said. I pulled my laptop off the shelf by my bed and, balancing it on my knees, I spent a few minutes connecting to the hospital X-ray system over the internet. I looked at the scan.

'He's not going to do well, is he?'

'No,' my registrar replied.

'Have you spoken to the family?'

'Not yet. He isn't married. There's a brother, who's coming in. Should be here soon.'

'What's the time?'

'Six.'

'Well, we can probably wait until the brother's here.'

Translated, the story – or 'history', as doctors call it – was of a young man who had suffered an intracerebral haemorrhage (ICH) from an AVM, an arterio-venous malformation. This is a kind of rare birthmark, a tangle of weak, abnormal blood vessels that often burst and bleed into the brain. The haemorrhage was into the left side of his brain, and also into part of the brain called the midbrain, which is important in keeping us conscious. It looked unlikely to me, from the scan, that if we operated he would get back to any kind of independent life. It is rarely possible to be certain, but I doubted if he would ever regain consciousness, let alone walk or talk again. His GCS was four, which meant he was in a deep coma. The scan showed a critical build-up of pressure in his head ('lots of shift on the CT', as my registrar put it). The fact that the pupil of his left eye was 'blown' – enlarged and no longer reacting to light – was a warning sign that without surgery he would probably die within the next few hours. The pupil had become smaller with a drug called mannitol, which temporarily reduces intracranial pressure, so we had a little time to decide what to do.

I couldn't get back to sleep and went into the hospital an hour later. The sun was rising over south London, a long line of bright orange seen through the hospital windows. The corridors were quiet and empty as it was so early in the morning, but the ITU was very busy and full of noise. The twelve beds were all occupied and the nursing shift was about to change over, so there were many staff milling around the nurses' station. There was a forest of drip-stands for intravenous fluids and syringe pumps, and flashing monitors standing guard beside each bed, the constant sound of the monitors bleeping and the softer, sighing sound of the ventilators doing the patients' breathing for them. The nurses were all talking, handing over their patients to each other.

The unconscious patients lay immobile, covered by white sheets, with tubes in their mouths connected to ventilators, IV lines in the veins of their arms, nasogastric tubes in their noses and catheters in their bladders. Some had drainage tubes and pressure-monitoring cables coming out of their heads.

My patient was in the far corner and there was a young man sitting at the bedside. I went up to him.

'Are you his brother?'

'Yes.'

'I'm Henry Marsh, the consultant responsible for Rob. Can we go and talk?'

We shook hands, and left Rob's bed to go to a small room used for interviews, for breaking bad news. I signalled to one of the nurses to join us. My registrar appeared, slightly out of breath.

'I didn't know you were going to come in so early,' he said.

I gestured to the patient's brother to sit down and sat opposite him.

'We need to have a very difficult conversation,' I said.

'Is it bad?' the brother asked, but he would have known already from my tone of voice that it was.

'He's suffered a major bleed into his brain.'

'The doctor here,' he said, pointing to the registrar, 'said you had to operate.'

'Well,' I replied, 'I'm afraid it's a bit more complicated than that.'

I went on to explain that if we operated and he survived, there was very little chance of his getting back to an independent life.

'You know him better than I do,' I said. 'Would he want to be disabled, in a wheelchair?'

'He loved the outdoor life, went sailing . . . had his own boat.'

'Are you close to him?'

'Yes. Our parents died when we were kids. We were best mates.'

'Girlfriend?'

'Not at the moment. Broke up recently.' He sat with his hands between his knees, looking at the floor.

We sat in silence for several minutes. It is very important not to try to fill these sad silences with talking too much. I find it very difficult, but have got a little better at it over the years.

'No chance?' his brother asked me after a while, looking up at me, into my eyes.

'I doubt it,' I replied. 'But to be honest you can never be entirely certain.'

There was another long silence.

'He'd hate to be disabled. He told me that once. He'd rather be dead.'

I said nothing.

'Rob was my best friend.'

'I think it's the right decision,' I said slowly, even though neither of us had explicitly stated what we had decided. 'If he was a member of my family, that's what I would want. I've seen so many people with terrible brain damage. It's not a good life.'

So the decision was made and we did not operate. Rob died later that day – at least, he became brain-dead, the ventilator was switched off and his organs were used for transplantation. I suppose it was just possible that I might have been wrong and he might have got back to some semblance of an independent life, or perhaps his brother was wrong, and Rob would have come to terms with being disabled, or simply have had no insight into it and led a happy, minimally conscious existence, no longer the person that he once was. Perhaps, perhaps – but doctors deal with probabilities, not certainties. Sometimes, if you are to make the right decision, you have to accept that you might be wrong. You may lose one patient with a good outcome but save a far greater number – and their families – from great suffering. It's a difficult truth that even now I find hard to accept. When I received phone calls at night about cases like this, if I told the surgeon on call in the hospital to operate, I would roll over and get back to sleep. If I told him not to operate, and that it was better to let the patient die, I would lie awake until it was time to go to work.

The operation on the six-year-old child in Nepal was only two days after the awake craniotomy and not especially difficult. I had done many similar cases before, but rarely with tumours of this size. It involved separating the two halves of the brain – the cerebral hemispheres – using what is known as a transcallosal approach. It was an operation I had always taken a particular interest in because it was the one that had saved my son's life many years earlier when, at the age of only three months, he had undergone surgery for a brain tumour. But his tumour had been only a fraction of the size of this one.

Dev and I had already decided that my principal role in his hospital would be to help him train his juniors so that they could learn how to do more than just the opening and closing and the emergency work at night. Within the first few days of coming to Nepal I knew, with the blind confidence of a lover, that as long as I could still usefully work, I would want to spend as much time as possible in the country.

'You can't go on running your hospital single-handed like this for ever,' I told him, 'and you need to think about the succession. You're not that much younger than me. What do you want to leave behind?'

'I know,' he said. 'It's been worrying me a lot of late.'

'I'd like to help, if I can be useful,' I went on. 'But you must promise to tell me as soon as I stop being useful.'

'I agree,' he replied.

You always get more nervous operating on children than on adults because of the terrible anxiety of the parents waiting outside the operating theatre. I had trained in paediatric neurosurgery and for many years had done most of the paediatric surgery in the hospital in Wimbledon. When that was closed and we were moved to the huge teaching hospital three miles away, I was unhappy about the way the paediatric neurosurgical ward was arranged. It was a very long distance from the neurosurgical theatres and my office, and it would no longer be possible for me to visit the children's ward several times a day as I had done in the past. It had been my way of coping with the parents' anxiety, which I understood all too well from the time when my own son had undergone surgery for his brain tumour and before I had become a neurosurgeon myself. To my shame and slight dismay, I found that I did not miss the paediatric work. In fact it was a relief to stop doing it.

'Can I have the microscope please?' I asked. The microscope – brand-new and as good as anything I have in London – was pushed into place.

'Has it been counterbalanced properly?' I asked. The second operation I had been involved with after arriving in Nepal had almost ended in disaster. The microscope, with an optical head weighing at least thirty kilos, had not been properly counterbalanced. The registrar had accidentally pressed the button that releases the perfectly floating optical head, and instead of floating it had crashed down onto my hands, almost forcing the instruments I was holding into the patient's brain.

With Dev's agreement we introduced a checklist to be completed before each operation, making sure (hopefully) that this would not happen again. I found it ironic that despite my well-known hatred of paperwork and checklists in my own hospital, I was now trying to introduce them in the hospital in Nepal.

'Yes, sir,' said Pankash, the registrar assistant, in answer to my question about counterbalancing the microscope. The registrars are very polite and respectful. If they do not know the answer to a question, they find it quite impossible to admit this. Rather than say no, they will stand speechless. The silence can last for many minutes and makes teaching very difficult. I had quickly resigned myself to never knowing what they really thought about what I told them or the questions I asked them.

I positioned the microscope and cautiously pressed the release button. The optical head remained steady and I settled down in the operating chair, which had also been pushed up to the table.

'What is the first rule of microsurgery?' I asked Pankash.

'To be comfortable, sir,' he replied. I had told him this the day before.

'Look,' I said. Pankash was peering down the microscope's side-arm so he could see what I was doing. Dev was watching on the monitor. I put the retractor against the inner side of the right hemisphere and gently pulled it a few millimetres to the right, away from the thick midline membrane called the falx cerebri that separates the two hemispheres. As I looked down the binocular microscope, it was as though

I was descending a ravine or negotiating a narrow crevice, with the shiny, silvery-grey surface of the falx to the left and the pale surface of the brain, etched with thousands of fine blood vessels, glittering in the microscope's brilliant light, to the right. Even after thirty years I still find using an operating microscope profoundly exhilarating – the feeling of beauty and mystery and exploration has never left me. After years of practice, the perfectly balanced instrument is like an extension of your own body and you feel – until things go wrong – equipped with superhuman powers.

'If we are lucky we will quickly drop right down onto the corpus callosum – there it is!'

The white corpus callosum came into view at the floor of the chasm, like a white beach between two cliffs. Running along it, like two rivers, were the anterior cerebral arteries, one on either side, bright red, pulsing gently with the heartbeat, which you must not damage under any circumstances. The corpus callosum contains countless millions of nerve fibres joining the two halves of the brain. If all of the corpus callosum is divided – as is occasionally done for severe epilepsy – patients develop what is called the 'split-brain phenomenon'. Outwardly they appear normal enough, but if placed in an experimental situation where the two cerebral hemispheres each see a different image, the two halves of the brain can end up disagreeing about what they are seeing – in particular, what something is called and what it is used for: knowledge of names is in the left hemisphere, and knowledge of how to use things is in the right. The self has been split. Actual conflicts between the two halves of the brain are rare, but it is said that one patient, on losing his temper with his wife, attacked her with his right hand while his left hand tried to restrain it.

Who has not felt contradictory impulses within themselves?

The more you learn about the brain, about our true selves, the stranger it becomes. It is almost as if we have many competing and cooperating selves within our brains, and yet somehow they all resonate together to produce a coherent individual capable of thought and action. There was a famous experiment many years ago by the American neuroscientist Benjamin Libet – confirmed many times since – that showed that the conscious decision to move the hand is *preceded* by electrical activity in the hand area of the brain. Nobody has yet provided a satisfactory account of what this really means. It is as though the deciding self is no different from the sailor in the storm who is forced to steer his boat in the storm's direction, but then claims to have chosen the direction himself. And yet to claim, as some do, that the conscious self is an illusion, that it is no more than running before the wind, or a consoling fairy story, somehow seems as implausible as maintaining that pain is an illusion, and not 'really' painful.

I was only going to make a small hole in the child's corpus callosum. This would bring me straight to the centre of her brain where the tumour was growing. A small callosotomy, as it is called, does not seem to produce any obvious problems for the patient after the operation. Besides, there were other, much greater risks to the child from what I was going to do, just as there were from not removing the tumour in the first place. And yet I was not at all certain that it was worth it: the child was almost certainly doomed whatever we did. Although the tumour was benign it was simply too large to remove completely without causing awful damage, both to the fornices and to an area near them called the hypothalamus. The hypothalamus controls vital functions such as thirst, appetite and growth. Children with hypothalamic damage typically become morbidly obese dwarves.

It was easy enough to find the tumour. It was at least four or five centimetres in diameter and of a soft, grey consistency which 'sucked easily', as neurosurgeons say. It was difficult to see if the all-important fornices still existed or whether they had been obliterated by the tumour. I managed to preserve a thin thread of white matter on one side that might, or might not, have been all that was left of them.

'Have a look,' I said to Dev, gesturing to the side-arm on the microscope. 'Do we know what her memory was like before the operation? It may already have been very poor, so there is less to be lost.'

'No,' he replied, 'but it's not going to make much difference to what we do, is it?'

'No, I suppose not,' I said, a little reluctant to accept his brutal realism. He scrubbed up and, once he had taken over from me, I went to have a cup of coffee.

Dev's office has a broad window looking out onto the space in front of the hospital, one floor down. In the distance you can see the green foothills to the north, which, whenever I look at them, invariably make me think of the celestial, snow-covered high Himalayas that lie hidden beyond them and which I long to see.

I drank my coffee and went back to the operating theatre. I scrubbed up and joined Dev and, after removing more of the tumour, we found that we had passed all the way through the child's brain onto the base of the skull on the right side. The tumour was so large that it had, effectively, split the lower part of the front of her brain in half.

'I think it's time to stop,' I said to Dev. 'We've taken out at least one fornix and one half of the hypothalamus – any more damage and she'll be completely wrecked.'

'I agree,' said Dev. 'She can have radiotherapy for what's left.'

'It's hard to know how much of the tumour is still there,' I said 'Maybe twenty per cent.'

There is no pleasure or glory in this kind of operating, I thought to myself as I tidied up the bleeding in the large cavity we had left in the girl's brain by removing the tumour. There's tumour left behind, she's almost certainly going to be left very damaged, and all we have done is slow down her dying.

6

THE MIND–BRAIN PROBLEM

'Patient is thirty-five-year-old man. He thinks there is insect in his head.'

'And you got an MRI scan?' I said.

'Yes, sir. No insect.' We looked at the scan.

'Well, you can tell him it's OK,' I said, though I had already seen so many cases of neurocysticercosis in the brain resulting in epilepsy or filariasis causing painful, swollen limbs and other problems that were entirely new to me that I had momentarily wondered if the man really did have some unusual skull-boring Nepali insect in his head.

'Shall we send him to see psychiatrist, sir?'

'Good idea,' I replied.

Once the day's operating is done the outpatient clinic is started. The patients will have been waiting all day, clerked by the juniors in the morning, and various investigations organized, and then seen by the more senior doctors, including the professor, once they have finished in the operating theatres.

I was ushered into the outpatient room on my first day to see a row of three patients and their families sitting next to the desk. In front of the desk stood five junior doctors. The patients looked startled and anxious. A receptionist brought some notes and one of the junior doctors, freshly out of a Chinese or Bangladeshi medical school, read out the history to me in stumbling but gabbled Nepali English, much of which I struggled to understand. The patient was an anxious-looking woman in a beautiful red dress.

'Patient is thirty-five-year-old and has headache for five years. Bowels and bladder normal. On examination pupils equal and reacting. Cranial nerves intact, reflexes equal and plantars downgoing. Had MRI scan.'

'Well, let's look at the MRI scan,' I suggested, which we did and which was, predictably, normal. How much does that cost? I wondered to myself. The answer, I learnt later, was an entire month's income. I was completely nonplussed. Uncertain as to what I was supposed to recommend, I asked the MOs.

After some hesitant discussion with them, I discovered that a huge variety of drugs were widely used in Nepal, often in a largely random manner. As it is, the patients can buy virtually any drugs themselves from small pharmacies on the streets. There is one on my walk to work, always with a queue. Steroids, I discovered, were popular for all manner of complaints, as was diazepam – Valium. After a few weeks of outpatient clinics, I began to suspect that the entire population of Nepal was on the pain-killing antidepressant amitriptyline.

The first patient was hustled off to be given a prescription and the next, who had been sitting next to her, was moved sideways onto the chair she had left. The clinic was clearly run on ergonomic, assembly-line principles. There was a long line of patients with headaches and backache, sore joints and one with rectal bleeding. I realized that the outpatient clinic functioned more as a GP surgery than a specialist neurosurgical clinic and I had to reach back into my basic medical knowledge from more than thirty years ago. This was both interesting – I was surprised at how much came back to me – and worrying. I was anxious that I might have forgotten something obvious and important after so many years spent specializing in neurosurgery. At least there was internet access, and it was helpful to find answers to most of my uncertainties on my laptop.

The next patient is a young woman with complete paralysis of half of her face after surgery for a huge acoustic tumour. It's a common complication and often inevitable if the tumour is as large as they usually are in Nepal because of delayed diagnosis. The patient and her husband are delighted when Dev comes into the room. They chatter happily. Dev puts his arm on the husband's shoulder.

'I was congratulating him on being a devoted husband. She was very ill after surgery but he stuck by her. They come from a part of the country where if the buffalo is ill, worth 63,000 rupees, they will spend money to treat it but not if the wife is ill. He's a good man!' And he slapped the man on the back again.

'Twenty-two-year-old woman with headache for three months. On examination pupils equal . . .'

'No, no, hang on a moment. What does she do for a living?'

There was a brief discussion between the MO and the patient.

'She counsels victims of torture, sir.'

'What? From the time of the Maoist insurgency?'

'Yes, sir.'

'Does she enjoy the work?'

Apparently she rather liked it. Had she received training for this? I asked.

'Yes,' came the reply.

'For how long?' I asked.

'Five days,' she said.

A skull X-ray was produced.

'This is a waste of time for headache,' I said.

'No, sir,' came the very polite reply. 'It is of her sinuses and she has sinusitis.' And now that I thought of it she certainly sounded as though she had a blocked nose.

'Ah, yes. I missed that. Shall we send her to the ENT clinic?'

'They are on holiday for *Dasain*, sir.'

'Well, you'd better prescribe her a decongestant then.'

And every so often there might be a patient with a brain tumour about whom Dev wanted my opinion, or another serious and often rare problem, but most of the patients had chronic headache or dizziness or the peculiarly Nepali symptom of total body-burning pain, and were determined to have MRI scans, despite my assurance that the scan would not help. As they would have to pay for the scan, it was not worth arguing over.

I quickly learnt that many of the patients were very disappointed to see me as opposed to the famous professor, even for the simplest of problems. I might have spent thirty minutes explaining things via one of the MOs but I had to resign myself to politely disappointed patients insisting on seeing him, although a few declared themselves happy with my opinion.

Meanwhile, in the room next door, Dev would be conducting his own high-speed clinic. The patients all expected to see him and he tried to see all the new ones himself. His room was full of doctors, receptionists and relatives, all standing, with the patient sitting in the middle of the melee. It made you think of a king, surrounded by courtiers and petitioners.

The door between our rooms was open and I could hear him coaxing, cajoling, declaiming, reassuring in rapid Nepali, depending on the class and education of the patients. They ranged from impoverished peasants from the mountains to teachers and politicians.

'How many patients actually have a neurosurgical problem?' I asked him.

'One point six per cent,' came the answer.

'Do other doctors refer you patients?'

'No, they all have their own connections and hate me. They try to refer them elsewhere but the patients come and see me anyway.'

As I left my first outpatient clinic I was stopped by a man I did not recognize.

'I am the girl's father,' he said in passable English. 'Thank you, sir, thank you so much,' pressing his hands together and holding them against his chest in Nepali greeting. Dev must have told him that I was involved in the surgery. I smiled, I hope not too sadly.

'My son had a brain tumour,' I told him, 'I know what you are feeling.' He thanked me profusely again, and I nodded in acknowledgement and sympathy and went to the management office to wait for Dev and to be driven home.

I have never enjoyed swimming – I was taught to swim at school at the age of eight, in the muddy river at the edge of the school playing fields, with a canvas belt around my waist attached to a rope and wooden pole, which one of the schoolmasters held like a heavy fishing rod. I dreaded having to climb down the slimy wooden ladder attached to the landing stage, with the cold, wet belt around me, seeing the master's shoes above me through the stage's planks, into the dark water. I would hang onto the ladder, half submerged, before being tugged by the master controlling the rope. I floundered into the water like a hooked fish. You were just expected to keep afloat by dog-paddling. There was no attempt to teach you to swim and the rope and pole were used to stop you drowning. I remember one of my schoolmates being flung into the river by the master when he was too frightened to descend the ladder. I used to wet myself with fear when changing into my swimming trunks for this character-building experience.

At my next school I was taught to swim properly by the kindly headmaster, but after that there was a notoriously sadistic ex-commando PE master who once hit my face so hard that it was numb for hours afterwards. I was so frightened of the man that I would slam my classroom desk's hinged lid on my hand to bruise it and claim that I had fallen and couldn't swim. That only worked once, so I then took to sticking my finger in one of my ears for many hours, mimicking an ear infection. The school doctor was very puzzled by this, as it only happened once a week. I was marched off to an ENT clinic at St Thomas's Hospital accompanied by the school matron. A sceptical consultant, with a row of medical students, looked in my ear and expressed some doubts. I can't remember what was said, but I do recall trying to persuade myself that there really was a problem with my ear even though I knew that I was malingering. It was my first experience of cognitive dissonance – entertaining entirely contradictory ideas – and the importance of self-deception in trying to deceive others. I then discovered that music lessons for playing the trumpet were on the same day and at the same time as the swimming class with the vile ex-commando, so I took up the trumpet but did not get on with it. Eventually I would just hide in a cupboard and skip the swimming lessons – an act of some bravery, I thought – and I got away with it.

I was at my weekly brain-tumour meeting twenty-five years later when a brain scan with a familiar name appeared on the screens in front of us. It was the PE master from my past and it showed a malignant brain tumour.

'He's a most unpleasant person,' my oncology colleague said. 'We've had no end of trouble with him but it's a frontal tumour, so maybe he's suffered personality change.'

'No, he hasn't,' I said, and explained my connection with the unfortunate man.

'The tumour needs to be biopsied,' my colleague said.

'I think it might be better if you got somebody else to do it,' I replied.

I wake with the dawn, the crack between the curtains facing my bed going from dark to light, to the sound of cocks crowing, dogs barking and birds singing. I go for a run every morning, but it took me a few weeks to overcome my fear of the local dogs – the guidebooks warn of rabies but my Nepali friends assured me this is more of a problem with the temple monkeys than the street dogs. So at first I ran in slightly absurd small circles and figures of eight in Dev and Madhu's garden, and up and down the many steps, for half an hour. Later, a little braver, I took to running for longer along the local lanes, between the tightly packed houses that didn't exist even ten years ago, past the rubbish and open drains, past sagging power and phone lines and bougainvillea hanging over garden walls. The road is uneven earth and rock, but there are a few short stretches of rough concrete, prettily patterned with the street dogs' pawprints. There is a small shrine on my usual route, and passers-by ring the bell that hangs by its entrance. All around me there is the sound of people coughing and hawking as they start the day. Neither the dogs nor the local people take any interest in me – it seems that there is nothing unusual in the sight of an elderly and breathless Englishman in football shorts stumbling along the road, but Nepalis are very polite and so perhaps are the dogs.

In England I run for longer. I used to run close on fifty miles a week, but one of my knees started to complain and now I only run twenty-five miles a week. I rarely enjoy it – I find it a considerable effort and my body feels stiff and leaden – but I do it for fear of old age and because exercise is supposed to postpone dementia. But there were occasionally wonderful moments when I was still running long distances – up to seventeen miles at weekends in the countryside surrounding Oxford. One early spring morning I was in Wytham Woods, the low sunlight falling diagonally through the trees, when I came across a leveret – a young hare – eating grass beside the path. It appeared completely unafraid of me and I was able to stand only three feet away as it quietly grazed, looking at me with its bright eyes. It was a unique moment of innocent trust from a wild animal, and I felt deeply moved. There is a beautiful ink and sepia drawing by the mystical early-nineteenth-century artist Samuel Palmer in the Ashmolean Museum in Oxford which shows the very same scene – a young hare in a wood, early in the morning, with the sun rising.

On another occasion, as I ran along the Thames, I noticed a duck desperately flapping in the water at the end of a broken-down pier. It appeared to be caught on something, so I crawled out along a steel beam projecting over the river, all that was left of the pier, feeling heroic. I found that the duck had a fish hook in its beak, with the fishing line wrapped around the beam. I managed to free it without falling into the river. The duck promptly dived into the water without stopping to thank me. Nevertheless, I like to think that if one day I ever get into trouble when swimming, the grateful duck – as in the fairy stories – will come and rescue me.

After running round Dev and Madhu's garden, I do fifty press-ups and a few other exercises, all of which I also hate doing, but I feel much better for it afterwards. I finish with a short swim in the small swimming pool outside the guest house. There is a very brief moment of ecstasy as I push out into the cold, mirror-calm water, which reflects the early-morning sky, with a view of the nearby Himalayan foothills in front of me. I momentarily forget my deep dislike of swimming. I complete this morning ritual with a cold shower – something I started doing two years ago. At first, admittedly in England in the winter, I thought I had discovered the elixir of life. A feeling of exhilaration, of intense well-being, would last for up to two hours afterwards. To my great disappointment, this wonderful feeling – acquired so easily within a couple of minutes – became shorter and shorter within a matter of weeks. I continue to have a cold shower every day, but the feeling now lasts only a few seconds at best, although the cold water still makes me jump about and gasp. I suppose my physiology has adapted, although health fanatics claim that cold is good for 'vagal tone' – the activity in the vagus nerve, which controls many of our body's functions in ways that we scarcely understand. It is a long nerve, which bypasses the spinal cord and reaches from the brain to the heart and many other organs, carrying information and instructions in both directions. It is an extraordinary nerve. Stimulation of the nerve with an electric current can help epilepsy, though nobody knows why. It can allow the generation of orgasms in women who are paralysed and have suffered complete destruction of the spinal cord. Apparently, people who have had it divided (an obsolete operation for gastric ulcer) will not develop Parkinson's disease.

After all this I sit beside the swimming pool in the little paradise of Dev and Madhu's garden, with flowers and birds all around me, and drink a cup of coffee before setting off for the hospital. Sometimes a bird with brilliant turquoise plumage dives down onto the surface of the pool, its wings and the splashing water flashing in the sunlight.

After a few weeks I decided to rearrange the way my clinic was run. I had the junior doctors sitting down, I would politely greet each patient when they entered, as I would do in England, which seemed to be less expected here. We would only have one patient in the room rather than a whole queue. The patients would usually come into the room looking expressionless, but my saying '*Namaste*' and pressing my hands together would almost invariably produce an utterly charming, slightly shy smile in reply. I insisted that every consultation had to end with asking the patient if they had any questions. This made the consultations feel a little less like assembly-line work but greatly reduced the number of patients I could see with the MOs, as the patients had so many questions to ask. They rarely spoke English and often were poor historians, as doctors call patients who have difficulties describing their symptoms. Many of them were subsistence farmers who could not read or write, and the MOs' English was often very limited as well. Making any kind of diagnosis could sometimes be impossible as the patients seemed so uncertain about their symptoms and were so determined to be given some new drug treatment. On the other hand, some of the patients had diseases such as TB and filariasis, with which I was unfamiliar. I found conducting the clinic extremely difficult, and had to be careful not to miss a serious problem in the constant stream of patients with chronic low back pain, headaches and total body-burning pain.

'Do you know what somatization is?'

'No, sir.'

'Well, it's the idea that if people are unhappy or depressed – marriage problems, things like that – rather than admit it to themselves, they develop headaches or total body pain, or strange burning feelings. They attribute their unhappiness to these symptoms, rather than consciously admit that they are unhappy in their marriage or that there is some similar problem. Such symptoms are called psychosomatic. You can see it as a sort of self-deception. Is the diagnosis of depression recognized here?'

'Not really, sir.'

'All pain is in the brain,' I explained as I pinched the little finger of my left hand in front of the MOs on the other side of the desk. 'The pain is not in the finger – it's in my brain. It's an illusion that the pain is in the finger. With psychosomatic symptoms, the pain is created by the brain without a stimulus from the peripheral nervous system. So the pain is perfectly real, but the treatment is different. But patients don't like being told this. They think they're being criticized.'

'Many of the women are seeking attention,' Upama, the MO said. 'Their husbands are away working abroad and they are unhappy.'

Amidst the flood of patients with minor problems, there are terrible cases as well – a young woman with much of her scalp infiltrated by a malignant skin tumour, a man dying from a brain tumour. There was a child, a thirteen-year-old girl, with half her face paralysed. The scan showed a complex congenital malformation of the joint between the spine and skull, which was the likely, though a very unusual, cause of her paralysis. Neither Dev nor I are very expert in such problems, and we had agreed that surgery was probably too difficult and dangerous. Upama explained this to the girl and her father, and the girl started sobbing silently.

'She is a girl,' Upama explained. 'Her face . . .'

While I watched the child cry, I thought about my detachment from her suffering – detachment both as a doctor and also because of the great gap of culture and language between us. I have to be detached, I thought, and it is something I learnt as soon as I qualified as a doctor. I cannot help this child, and there is little point in being emotional about it. But I also thought of the research into bonobos (previously known as pygmy chimpanzees), our closest evolutionary relatives, which shows that they have compassion and kindness, a sense of fairness and console each other over pain – at least for their own group. They have not been told to do this by priests or philosophers or teachers, it is part of their genetic nature, and it is reasonable to conclude that the same applies to us.

For most of us, when we become doctors, we have to suppress our natural empathy if we are to function effectively. Empathy is not something we have to learn – it is something we have to unlearn. Patients become part of the 'out-group' as anthropologists call it, people with whom we need no longer identify. But the child went on crying and I started to feel uncomfortable. Besides, I told myself, the only way that doctors can lay claim to any kind of moral superiority over other professions is that we treat – at least in theory – all our patients in the same way, irrespective of class or race or nationality, or even of wealth. So my detachment wilted as the child cried and I thought I might just see if Dev and I could be wrong. I used my smartphone to photograph the girl's scans and emailed them to a colleague on the other side of the world, an expert in problems of this sort, for an opinion. He replied thirty minutes later, saying he felt that surgery was both possible and relatively straightforward.

I showed his opinion to Dev.

'Isn't the internet wonderful!' I said. 'We can get a world-class opinion so quickly.'

'We'd better get the child back and talk to the family,' he replied, but the girl and her family had disappeared.

While the patients come and go, the day outside grows dark. The high Himalayan foothills on the horizon disappear. The ragged leaves of the banana tree in the paddy field next to the hospital start to shake and flap in the wind. A flock of small birds is suddenly flung up into the sky like a handful of leaves, to be quickly swept from sight. The windows of the outpatient room are open – the room fills with the intoxicating smell of wet earth and the patients' notes in front of me blow off the table. There are frequent power cuts and every so often the room plunges into darkness for a few minutes. Thunder crashes directly overhead, to echo away into the distance.

'Patient is sixty-five-year-old man with numbness in his fingers.'

The MRI scan shows slight compression of the sixth cervical nerve roots in his neck.

'How much is he troubled by his symptoms?'

'He has difficulty climbing trees and milking buffalo, sir.'

We decide to continue with conservative treatment.

'It is proxy case, sir. Father has brought scan. His two-month-old daughter is in other hospital. They have diagnosed bacterial encephalomeningitis. Child is fitting, and they grew enterobacter in the blood. They have recommended three weeks of IV antibiotics. He wants to know if the treatment is right.'

The CT scan was of poor quality and I found it difficult to interpret, but it looked as though the child might have suffered extensive brain damage.

'He wants to know if it is good idea to spend money treating the child.'

'How many other children does he have?'

'Three, sir.' But then we worked out that two of them had already died.

I looked at the scan for a long time, not knowing what to recommend.

'I think I'd get an MRI scan,' I eventually said. 'If it shows severe brain damage, perhaps it is better to let the child die.'

Jaman, the excellent MO, spoke to the father.

'It is economic problem for the MRI scan,' he told me.

'Then it's very difficult,' I said.

I left Jaman and the other MOs to have a long conversation with the father. I don't know what was decided, but the father said '*Namaste*' to me very politely as he left.

'Patient is forty-year-old lady who has had headache for twenty years, sir.'

My heart sinks a little.

'Well, tell me more about the headache.'

We discuss this for a few minutes. The patient has been on a long list of drugs over the years.

'She suffers from panic attacks. She finds diazepam helps, sir.'

I deliver a little lecture on the evils of diazepam and the way that millions of housewives became addicted to it in the past in Europe and America. It is very difficult to know what to suggest.

'Do you know the word stigma?'

'Yes, sir.'

'Is there stigma in Nepal against seeing psychiatrists?'

'Yes, there is, sir.'

'I think you should suggest she sees a psychiatrist. I find it helps if I tell patients that I had psychiatric treatment myself once. It was invaluable.'

There was a rapid exchange in Nepali.

'She wants MRI scan, sir.'

'It's a waste of her money.'

'But she lives in Nepalgunj.'

'How far away is that?'

'Two days by bad road.'

'Oh all right, get an MRI scan then . . . it won't show anything but I suppose she hopes that somehow it will make her unhappiness real.'

Afterwards the MO tells me that the patient has already tried to kill herself twice.

'How do people kill themselves in Nepal?'

'Usually by hanging, sir.'

The patients come from all over Nepal, often from remote mountain villages accessible only on foot. They come to the clinic hoping for an instant cure, and with an exaggerated faith in medicines, perhaps connected to their belief in prayer and sacrifice. The idea that drugs can have side effects, that there is a balance to be struck between cost and benefit, seems very alien to them. It is impossible to treat effectively chronic problems such as headache, epilepsy, raised blood pressure or low back pain on the basis of a single visit. So the patients end up on a bewildering variety of different drugs that they either acquire themselves or from different doctors over the years. They come with plastic bags full of shiny foil blister packs of coloured tablets of many shapes and sizes, which they spread out on the table in front of me and the MOs.

'She is thirty-year-old lady with headache, sir.'

Oh dear, I thought, not another one. She sat diffidently in front of me with her husband beside her.

'And she cannot stop laughing, sir.'

'Really? Pathological laughter? That's interesting.'

I was handed the scan. It was indeed very interesting, but very sad.

'What do you see, Salima?'

After a while, Salima, with my help, worked out that we were looking at a huge brain tumour – technically a petroclival meningioma. I had once had a similar case in London who also had the very rare symptom of uncontrollable, pathological laughter. I had operated, and had left her in a persistent vegetative state. It was one of the larger headstones in my inner cemetery.

'Tell them to come back tomorrow when Prof is here,' I said.

Once she had left the room I told the MOs that without surgery the poor young woman would die within a matter of years – slowly, probably from aspiration pneumonia. She already had difficulties swallowing, from the pressure of the tumour on the cranial nerves that controlled her throat, a sure harbinger of death. But surgery, I told them, was almost impossibly difficult – at least, it was very difficult to operate without, at best, inflicting lifelong disability on the patient. So what was better? To die within the next few years, or face a longer life of awful disability?

'Prof needs to talk to them,' I said, but she never came back.

'All is well apart from the child . . . the baby where we tried to do an endoscopic ventriculostomy yesterday.' This had been another patient, a baby only a few months old, with a huge hydrocephalic head.

'In what way?'

'Not doing well . . .'

'Does the mother have other children?'

'Yes.'

'It's best if we let her die, isn't it?'

Dev said nothing but silently conveyed his agreement.

'In England we wouldn't be allowed to do that,' I said. 'We'd raise heaven and earth and spend a fortune to keep the child alive even though she will have a miserable future with severe brain damage and a head the size of a football. My old boss, at the children's hospital where I trained, sometimes said to me, after we had operated on a particularly hopeless case who was doomed to live a miserable and disabled life, that he wished he could tell the parents to let the child die and go and have another one. But you're not allowed to say that.'

'The child died during the night,' the registrar told me when he saw me next morning looking at the space where she had been. The child had gone, leaving only a sad little huddle of sheets on the bed, as the nurses had not yet had time to change the bedding.

I had some difficulties setting up a patient for an MVD, an operation for facial pain which involves microscopically manipulating a small artery off the trigeminal nerve, the nerve for sensation over the face. It is an operation I have done hundreds of times in London – but doing it here feels very different. Turning the patient was somewhat problematic.

'In London we say one, two, three and then turn the patient,' I said. 'Do you do that here?'

'Yes, sir,' the registrar assured me happily.

'One,' I said, and he grabbed the patient and started pushing him off the trolley.

'No! No!' I shouted. 'One, two, three . . . and *then* roll.'

It felt more like a rugger scrum than a coordinated manoeuvre, but we managed to get the patient safely face-down onto the table.

*

It was a twenty-minute drive from Neuro Hospital to the Bir, past a few small demonstrations with heavily armed police in attendance. Nepal is in a constant state of political chaos. The civil war only ended a few years ago. The monarchy collapsed four years after the royal massacre. The democratically elected Marxist government which replaced it is riven by continuous political infighting. The streets around the hospital were packed with pedestrians and motorbikes. An emaciated young woman was selling a few halved cucumbers, daubed with a red relish, from an empty oil drum that served as a stall, at the hospital entrance. There was a row of ramshackle pharmacies across the road from the entrance, with crowds of people standing in front of them.

'That was the first pharmacy in Nepal,' Dev said, pointing to an old brick building behind the pharmacy shacks with wide cracks in its walls from the recent earthquake.

The hospital itself was more like a dirty old warehouse. It reminded me of some of the worst hospitals I had seen in Africa and rural Ukraine. It had been built in the 1960s by the Americans, and although some of the wards had plenty of windows, it was a typical example of the style of architecture that treats hospitals as being little different from factories or prisons, with long, dark corridors and lots of gloomy spaces. The wards were very crowded and everything felt sad and neglected. Dev was greeted with many delighted smiles and '*Namastes*' by the staff, but he told me afterwards that he had been deeply upset by the visit.

'I created my own neurosurgical unit here,' he told me. 'The first in Nepal. We had to build everything from scratch with second-hand equipment. I used to do my own cerebral angiograms by direct carotid puncture in the neck. Jamie Ambrose at AMH had shown me how to do it. We painted the ward every year – I paid for the paint myself – we had a painting party. And look at it all now! All gone, filthy, neglected.'

'When I came back here from the UK,' he continued, 'nobody would work after two in the afternoon. So I sat in the office by myself, the only senior doctor in the building. Eventually everybody else stayed as well. We had no money then. I was working all the time.'

We left the hospital and stood outside, waiting for Dev's driver. Dev was recognized by many people – he is famous throughout Nepal, let alone at the hospital where he used to work – and while he chatted and joshed with them I stood and watched the endless flow of people coming and going. There was a large pool of dirty water from a leaking water main, and rubbish and broken bricks – probably left over from the earthquake – in the gutter opposite. And yet, as the women picked their way across the road in their brilliantly coloured and elegant clothes, I thought, with a slight feeling of shame, that the scene was rather beautiful.

As Ramesh, Dev's driver, manoeuvred the car past the long and chaotic queues outside the petrol stations, Dev returned to the subject of the Bir.

'I need a rest after what I have just been through. It was terrible, terrible . . . people would come to appreciate just how good a ward could be . . . all gone. That floor was something different. Nice working environment. It was recognized by the Royal College in England for training. All gone, all gone.'

A few months later I met an English neurosurgeon in New Zealand who, when a medical student, had visited Dev's department at the Bir. He fully confirmed just how different the department had been from the rest of the hospital.

'It was a beacon of light in the darkness,' he said.

'We came back here with such high hopes,' Madhu told me over supper that evening, 'and everything has got so much worse.'

7

AN ELEPHANT RIDE

We set off for the Terai – the flat lowlands in the south of Nepal, bordering India – early in the morning, just after dawn, the air very still and humid and hot. I was in a cheerful mood: the day before, I had received the pathology report on the skin tumour I had had removed before setting out for Nepal. The tumour was indeed cancerous but the 'excision margins were clear' – in other words, I had been cured and would not need further treatment.

There is an entire tourist village centred on elephant rides adjacent to the Chitwan National Park. Tourism in Nepal had suffered badly because of last year's earthquake and the nearby tourist town of Sauraha, with many bars and small hotels, looked almost empty when we drove through it.

'What will they live off?' I asked Dev.

'Hope,' he said, with a shrug.

There were only a few Western people to be seen, easily identified by their baggy shorts and T-shirts. I always wear a long-sleeved shirt and trousers, not just to be different but because the Catholic missionaries I lived with in Africa fifty years ago, when I was working as a volunteer teacher, taught me that this shows respect for the local people.

We were taken to the government elephant station at the edge of the jungle. We walked beneath tall, widely spaced trees, through patches of sunlight. It was remarkably quiet. There was a group of elephant-high, ramshackle shelters – roofs of battered and rusty corrugated-iron sheets on four posts, surrounded by tattered electric fencing. In the centre of each shelter there was a massive wooden pillar, with heavy chains and shackles hanging down. There were no elephants to be seen.

'They used to keep the elephants chained at night but an Englishman showed them they could use electric fencing,' Dev told me.

Beyond the sheds were a few low buildings, and in one with an open front two European teenage girls were sitting cross-legged in very short shorts on the ground next to a dark-skinned elderly Nepali man. They were rolling up handfuls of rice mixed with sweets – the plastic wrappers were carefully removed – into a ball, wound around with long grass, a packed lunch for elephants. They held the ball with one foot and used their hands to bind the long grass around the rice. The girls were very absorbed and everybody was silent. When I asked them where they came from, they smiled and said they came from Germany. I wasn't quite sure what to think about seeing these children of the affluent West playing at being peasants.

And then, slowly, coming out of the surrounding jungle, a huge elephant appeared with a mahout perched high on her neck, his feet behind her ears. The creature was enormous, solemn and stately, and strangely graceful for such a massive beast. The last of the megafauna on land to survive mankind.

'That's the one we'll be going on,' Dev told me.

The mahout brought the great creature to where we were waiting, and the elephant bent her huge knees and settled awkwardly down on the ground, back legs pointing backwards and front legs forward. The mahout and his helpers then spent some time fitting a wooden frame over a mattress onto the elephant's back, kept in place with a wide girth, which they heaved on with ropes to get it tight. While they did this I walked up to the elephant and looked into her small, thoughtful eyes and she looked back at me. She had elegantly curled the end of her trunk over her left foreleg. The day before I had been reading about elephants – of the 40,000 muscles in their trunks, and of their great brains, the largest brain of all land mammals. They are intensely social, with a complex social life. They can console each other, mourn the dead and have a language of sorts. They can also recognize themselves in mirrors (which is generally considered to mean that they have a sense of self).

Nobody knows how many brain cells are needed for consciousness. Recent work on insects suggests that even they might be capable of it; their brains show similarities to the midbrain of reptiles and mammals, where some authorities think conscious experience arises. To ask if a creature is conscious is equivalent to asking if it can feel pain, and nobody knows at what point pain arises in nervous systems. If you deliver a painful stimulus to one of a lobster's claws, it will rub the painful area with the other claw. Is this a mere reflex? It seems more likely that it feels pain. We boil lobsters alive, of course, before eating them.

When patients are unconscious, for instance after a head injury, we assess the depth of their coma by hurting them. You either squeeze the nail bed of one of their fingers with a pencil, or press very hard with your thumb over the supraorbital nerve just above one of the eyes. If they respond purposefully to the pain – trying to push you away or, just like the lobster, trying to get one of their hands to the painful area – we assume there is some kind of conscious perception of pain going on, even if the patient has no memory of it afterwards. On the other hand, if the patient is in deep coma, they do not move in response to the pain at all, or move their limbs in a reflex, purposeless way. We assume, then, that there is no conscious element to the response and that the patient is deeply unconscious.

And then there is the wonderful mystery, at the other end of the scale from insects, as to why whales have brains which are so much larger than ours. It is true that there are structural differences (whales lack our cortical layer IV and most of them have a much higher ratio of supportive glial cells to neuronal cells than we do), but nobody knows why they have evolved such massive brains, and to what use they are put. In recent years the floodgates have opened with new research into animal intelligence: cows have friends among other cows, pilot whales (a species of dolphin) have more neuronal cells in their brains than any other creature, manta rays can recognize themselves in mirrors, fish can communicate and work together when hunting. We are moving further and further away from Descartes's separation of mind from matter, and his dreadful claim that animals are mere automata.

Self-consciousness, however, to be aware of one's own awareness, to think about thinking, is probably a more complex phenomenon. I first discovered it at the age of fourteen on a school expedition to the ruins of Battle Abbey on the South Coast. With the other boys, I ended up larking about on the nearby shingle beach. I ran fully clothed into the sea and stood with the waves lapping about my knees, soaking my school uniform. As I stood there, I was suddenly struck by an overwhelming awareness of myself and of my own consciousness. It was like looking into a bottomless well, or seeing myself between a pair of parallel mirrors, and I was terrified. We returned to London in a coach and I came home in a state of considerable distress. I tried to explain what I felt to my father as he sat in his book-lined study. I started to shout about suicide, but I think he was rather confused by my hysterical outburst, as was I.

Clearly this sudden self-awareness was a philosophical version of the awkward self-consciousness which comes to boys with adolescence, when testosterone levels rise. I remember the shock I experienced on seeing my first, solitary pubic hair. Over the next two years I had a series of what are best described as mystical experiences – feelings of profound illumination and unity associated with intense visual effects, where shadows and colours acquired an extraordinary depth and beauty. My hands, and the veins on them, would look especially profound. I would gaze at them with wonder.

When I was a medical student many years later, studying anatomy, I was particularly fascinated by the anatomy of the human hand. There was a large polythene bag in the Long Room – the room with corpses for dissection – full of amputated hands in various degrees of dissection. The hand is a wonderfully complex mechanism, with a series of tendons and joints and muscles, a device of articulated levers and pulleys. I drew and painted careful and elaborate watercolour studies of these hands, but to my regret lost my anatomy notebooks many years ago. I subsequently discovered in Aldous Huxley's writings that my mystical experiences were identical to those he described while taking mescaline. There is a form of epilepsy, known as limbic epilepsy – Dostoevsky is thought to have had it – where people have an intense feeling of unity and transcendence, and often interpret it as being in the presence of God. The limbic system is part of the human brain involved in emotions, and in 'lower' mammals is mainly involved in olfaction – the sense of smell. When I was at Oxford University most of my friends were experimenting with LSD, but I never dared. I smoked cannabis occasionally but disliked the complacency it produced.

The mystical experiences faded as I grew older, replaced, perhaps, by sexual desire and sexual anxiety. While my contemporaries at school were going to parties and learning to kiss girls, I sat in my room at the top of the large house in Clapham, reading voraciously. I kept a diary which I destroyed in a fit of embarrassment and shame a few years later. I rather regret that: I suspect that many of the questions and problems that trouble me now, as I face retirement and old age, were already present then, when I was also trying to find a sense of purpose in my life, but had much more of it ahead of me. It would also amuse me to see what a prune I was when young, and how seriously I took myself.

My father recommended many books, ranging from Raymond Chandler to Karl Popper's *Open Society and Its Enemies* – this latter book, I think, had a great influence on my later life. Popper taught me to distrust unquestioned authority, and that our moral duty in life is *to reduce suffering*, by 'piecemeal social engineering' and not with grand schemes driven by ideology. This, of course, is very close to the Christian ethics and belief in social justice inculcated in me by my parents, and the understanding of the importance of evidence and honesty that I learnt as a doctor. Yet doctors get paid – usually very well – for their work, and we cannot but help people (unless singularly incompetent). So our work need not call for any particular moral effort on our part. It is easy for us to become complacent, the worst of all medical sins. The moral challenge is to treat patients as we would wish to be treated ourselves, to counterbalance with professional care and kindness the emotional detachment we require to get the work done. The problem is to find the correct balance between compassion and detachment. It is not easy. When faced with an unending queue of patients, so often with problems that we cannot help, it is remarkably difficult.

My experience as a hospital operating theatre porter had resulted in my deciding to become a surgeon. I had returned to Oxford to complete my degree before trying to get into medical school. My arrival back at Oxford was shortly followed by my first, and unsuccessful, attempt at sexual intercourse (with a sweet girl in Leicester who took pity on me). This precipitated a further crisis. I started to suffer from manic pressure of thought, seeing all sorts of wonderful connections between disparate things – at first rather exciting, but then very frightening. My ideas started to spin out of control, and the feeling of brilliant omniscience was replaced by a fear that there was some kind of evil presence beside me. I can see now that part of me was trying to force another part of me, through fear, to seek help. There is another form of limbic epilepsy, it is interesting to note, where people experience the presence of evil rather than the presence of God. At a friend's suggestion – another person to whom I am deeply indebted – I got in touch with the psychiatrist my father had unsuccessfully tried to persuade me to see a year earlier, when I had abandoned my degree. I was admitted for a short time to a psychiatric hospital.

I had a room to myself and lay there on the first night feeling miserable and tense. A friendly West Indian nurse came in and asked me if I wanted a sleeping pill.

'No, I don't need one,' I said defensively.

'Well, I'm called Charlie and I'm at the end of the corridor if you change your mind,' he said with smile.

I could not sleep. I had sunk so low that I had no future whatsoever. I had reached the bottom of a bottomless well, and there was no way up again. I had become a mental patient. I was utterly and completely alone. I cried and cried, but even as I cried I felt something frozen in my heart thawing, just like the fragments of the evil magician's mirror in the boy's heart in Hans Christian Andersen's story *The Snow Queen*. I had been fighting myself for so long, and for so long I had viewed other people only as mirrors in which I tried to see my own reflection (I am, alas, still prone to this). Was it that I had tried to freeze my heart, trying to suppress my hopeless and inappropriate love for the woman who had kissed me? I do not know, but I got out of bed in the early hours and walked down the dark hospital corridor to where Charlie was reading a newspaper, spread out on the table in front of him, in a little pool of light from a desk lamp. I asked him for a sleeping pill – it was Mogadon in those days, now banned. I fell asleep exhausted, and next morning I was rather pleased to see in the bathroom mirror that my inner suffering had finally become real, at least visible. Much better than trying to lacerate my hand and its veins with broken glass, I now had two enormous purple bruises under my eyes.

The next week I spent an hour a day with a deeply sympathetic elderly psychiatrist, unburdening myself. The feeling of being reborn, of being in love with everybody and everything, was intense. When I was discharged I drove up into the Chiltern Hills. It was a perfect autumnal afternoon. My body felt so stiff that it was as though I had run a marathon. I remember the difficulty I had climbing over a padlocked field gate. It was the happiest day of my life.

Research has shown that the ecstasy of being in love rarely lasts more than six months. It fades, and comes to be replaced by the more mundane practicalities of maintaining a successful relationship, but at least it fades a lot more slowly than the ecstasy I felt when I started to take cold showers. The intense feeling of illumination and optimism, of being part of a coherent whole that I felt after leaving hospital, was identical to many accounts I have read of religious conversion and revelation, except that I did not for a moment believe in any kind of divine presence in my life, or in the world. These intense feelings obviously involve the same cerebral mechanisms as when you are in love with a person, with the feeling of joyful unity, beauty and coherence all focused on that one person.

Zebra finches and other birds can grow new brain cells when the mating season begins, when they need to start singing to attract a mate. I wonder whether similar processes go on in our brains when we are in love. I also wonder whether other animals experience ecstasy. It has been suggested that the huge brains of dolphins and whales, creatures who also show great playfulness, mean that they do. It is easy to believe this if you watch a pod of dolphins swimming and leaping through the sea. I did not find God with my ecstatic experience, but instead I learnt that my own mind was a profound mystery, and that the sacred and the profane are inextricably linked. There must be a neural correlate for this, whereby the deep and basic instinct to procreate, present in almost all living things, becomes interwoven with the complex feelings and abstract reasoning of which our larger brains have evolved to be capable. This feeling of mystery about my own consciousness, but without any ecstasy, has grown stronger and stronger in recent years as my life starts to unwind and descend to its close. It is, I suppose, a substitute for religious faith and, in part, a preparation for death.

On one of my trips to the Sudan I had been taken to a small zoo in a huge sugar plantation in the desert, on the banks of the White Nile, a few hundred miles to the south of

Khartoum. There was an enclosure with five Nile crocodiles, who eyed me thoughtfully – they prey on humans – half submerged in their concrete pond. Next to it was a cage with a single young elephant in it. Deprived of its mother and its highly developed social life, it had clearly gone mad, and showed the same disturbed behaviour of grotesque and repetitive movements as severely autistic children, or the chronic schizophrenics I once cared for as a nursing assistant. And next to the poor elephant's cage there was a small enclosure with a young chimpanzee who seemed to have gone as mad as the young elephant. My Sudanese colleague – whom I greatly liked – roared with laughter when he saw my dismay.

'You English! You are so tender-hearted!' he said.

The look the Nepali elephant gave me, with her small, red-rimmed eyes – or so it seemed to me – as the girth was tightened round her was of deep and infinitely sad resignation.

We were taken to a twelve-foot-high mounting platform, with a staircase of rotten treads, overgrown with moss and climbing plants. The elephant was positioned alongside it, and Dev and I and two guides clambered into the wooden frame on her back, each of us sitting in a corner, facing outwards, our legs straddling one of the four corner posts. There was a thin cushion and it was less uncomfortable than I expected.

It was quite disconcerting at first, the slow, rocking movement, twelve feet off the ground, with the elephant gently placing her huge feet on the uneven track as we headed back into the jungle. This is going to be a bit boring, I thought, once I had got used to the swaying motion – there is nothing to do. But after a while I started to enjoy it, although I still wondered what the elephant thought about it all.

The mahout carried a sickle and a stick, and he used the

sickle from time to time to clear the way, as did the elephant with her trunk, coiling it expertly around branches and snapping them off. The sickle, I had read, could also be used to cut the elephant's ears if it became difficult to control. I have also read that training young elephants can involve considerable cruelty, although people also write of the close relationship between the mahouts and their elephants and of the benefit to conservation from the tourist income generated by the elephant rides.

Apparently you cannot get an elephant to do something it doesn't want to do, and watching the mahout and the animal as they chose which way to go through the dense jungle, it was clear some kind of negotiation was continuously going on. The mahout used his feet to kick gently behind her ears to steer her, like the pilot of an airplane using a rudder bar, but it was obvious that the elephant did not always agree with his suggestions. We crossed a river, the elephant effortlessly climbing the steep bank opposite, and went deeper and deeper into the tangled trees of the jungle, along paths that were almost invisible. In a small clearing we saw a herd of spotted deer which took fright and quickly disappeared, bounding with great elegance back into the trees. Apparently there are tigers and leopards in the reservation as well, but they are rarely seen. After an hour of this slow, rhythmic plodding between trees, the leaves brushing our faces, we emerged into grassland, some of the grasses almost as tall as the elephant. The mahout pointed out an area of flattened grass and said something to Dev.

'Rhino's bedroom,' Dev translated, and shortly afterwards, near the riverbank, we came across a rhino with a young calf, which quickly hid behind its mother as the monstrous shape of the great elephant with five human beings on top of it approached. Its mother took little notice of us, continuing to graze, as we admired the studded armour-plating of her skin and her single horn which the Chinese and Vietnamese so stupidly prize, ground up as an aphrodisiac or as a cure for cancer, resulting in the near-extinction of the creature from poaching.

'Why can't they use Viagra?' I complained, as we left the rhino and her calf behind and crossed back over the river. 'I'm sure it's cheaper.'

With Dev as translator, I asked the mahout about the elephant as we plodded majestically through the tall grass. He told me that she was forty-five years old and would probably live to seventy, but recently they had lost several elephants to TB. This elephant had had four calves, but three had died before they were three years old.

'When are the calves taken from their mothers to be trained?' I asked.

'At three years old,' I was told.

I asked if all the elephants were kept alone and was told that they were. As we crossed back over the river on the return journey the elephant suddenly let out a great trumpeting cry.

'What was that about?' I asked.

'She smells another elephant,' Dev translated.

Back at the elephant station, we climbed off our elephant's back and had to wait for a while for Dev's driver and bodyguard to appear. We sat in the sunshine outside a group of huts which clearly had been financed by some well-meaning foreign charity – a lopsided and mildewed notice announced that this was the Children and Women Promotion Centre. The notice was so faded that it was difficult to read, but there was a long list of projects and among these I could just make out 'Computer Letchur' (*sic*), 'Sports Coachers (any)', 'Environment', 'Caring Wildlife (injured)' and 'orphan animal',

'HIV/Aids awareness programme' and other projects funded by foreign aid. 'Unskill volunteers' were accepted. There was another notice, also worn and partly illegible, announcing a Vulture Recovery Programme, with the icons of all manner of international bird charities at the foot of it. The buildings were all dilapidated, with rusty corrugated-iron roofs. The shop was almost empty apart from some cheap imports from China and a single woman in attendance who, most unusually for Nepal, did not smile when I entered. All the world wants to help Nepal and vast sums of aid have been lavished on the country, yet much of it seems to have disappeared without trace, leaving only faded signs and notice-boards behind.

I cheerfully volunteered to join Dev operating on a large brain tumour in an eight-year-old boy and was soon regretting it. The tumour bled like a stuck pig right from the start and there were extraordinarily large arterialized veins running in it that bled furiously and too heavily for the diathermy to work. I started sweating. The problem is that when you worry about the patient bleeding to death you rely on a close working relationship with the anaesthetist, and she didn't speak English and was very uncommunicative. As I struggled to stop the child bleeding to death from a blood vessel in the centre of the tumour, I despaired of ever managing to train Dev's juniors to do such operating. It becomes almost impossible if you are trying to train your junior and you have to watch passively while they fumble and stumble, putting the patient's life at risk. It is easy to see why so often trainees are left to operate on their own, learning the hard way, on the poor and the destitute, who are unlikely to complain if things go badly. In all the countries where I have worked over the years, people with money or influence will make sure that they are not trained upon.

In poor countries such as Sudan and Nepal, there has been an explosion of private clinics and hospitals. The professional associations, largely based on the old British model, have become sidelined, and there is less and less effective maintenance of professional standards. Money and medicine have always gone together: what could be more precious than health? But patients are infinitely vulnerable, from both ignorance and fear, and doctors and health-care providers are easily corrupted by profit-seeking. It is true that socialized health care, as the Americans call it, has many faults. It tends to be slow and bureaucratic, patients can become mere items on an impersonal assembly line, clinical staff have little incentive to behave well and can grow complacent. It is often starved of resources. But these faults can be overcome if high morale and professional standards are maintained, if the correct balance between clinical freedom and regulation is found, and if politicians are brave enough to raise taxes. The faults of socialized health care are ultimately less than the extravagance, inequality, excessive treatment and dishonesty that so often come with competitive private health care.

Dev took over and I was able to go and have a sandwich. In fact the worst of the bleeding had stopped by then, but it was rather wonderful to be able to pause and have a break. And I thought, imagine running a practice like this single-handed for thirty years, with nobody to help out or relieve you – day in, day out and on call every night.

When I saw the boy on the ITU round next day he was awake and crying, and at first I thought that all was well. But something troubled me: his eyes were open, roaming and unfocused. I had missed it at first, but when I came back to see him after looking at the other patients it was clear that he was completely blind.

'What was his eyesight like before surgery?' I asked Dev.

'Not good,' he replied.

'He must have had severe papilloedema from the severe hydrocephalus,' I said. 'We know that some of them wake up blind after surgery.'

When I saw Dev later in the day he told me that he had seen the boy's mother.

'She said his eyesight was very bad indeed before the operation.'

'I have seen that happen twice before,' I said, 'it can't be avoided.' It was better not to think of the boy's future.

One of the first operations I did on a child, when I was a junior registrar, was on a nine-year-old boy with an acute subdural haematoma – a severe traumatic head injury – from a car crash. A neighbour was taking him with his own children to the zoo when another car drove into them. The neighbour was killed, as was his five-year-old daughter. The brain of the child I operated on became so swollen during the operation that I could scarcely get it back into his skull and I even had difficulties stitching the scalp back together over it. This can happen occasionally with acute subdurals. It turned out he was an only child, conceived after years of fertility and IVF treatment. There was no question of the mother having any further children. I had to tell his mother that he was going to die. I watched her as I told her this and realized I was delivering a death sentence, on her as much as on her child. It is not a good feeling to destroy somebody like this. The hospital where I was training was a high-rise building in the north of London, and the ITU had large windows with panoramic views of the city below. I remember how the light from the windows was brightly reflected on the ITU's polished floor as I led the mother to her son's bed, where he lay on a ventilator, a large, lopsided bandage hiding my rough stitching. I thought of how difficult it is to believe in a benign deity intervening in human life when you have to witness suffering like this. Unless, of course, in the words of the famous Victorian hymn, there really is a friend for little children, above the bright-blue sky, who will right all the wrongs we suffer in this life on earth in an afterlife in heaven.

But I am a neurosurgeon. I frequently see people whose fundamental moral and social nature has been changed for the worse, often grotesquely so, by physical damage to the frontal lobes of their brains. It is hard to believe in an immortal soul, and any life after death, when you see such things.

The outpatient clinic is usually finished by six in the evening. Dev's bodyguard will always materialize, as if by magic, at exactly the right time, and we are driven the short distance home. We then sit in the garden, drink beer and talk.

The kidnappers who had kidnapped their daughter six years ago had come up from the valley, poisoning one of Madhu's dogs with a piece of meat they threw over the fence, before climbing over the spiked fencing that surrounds the garden.

'It wasn't just being held at gunpoint and having your daughter kidnapped. There were extortion attempts as well. I used to carry my own mobile phone until I was telephoned one day. "You have heard of the Black Spider group?" the voice said. "You remember how we killed Dr So and So?" They wanted money but I took no notice, and now my driver Ramesh always carries my phone. And during the Maoist insurgency the Maoists often came demanding money. I always refused but said I was happy to give them free medical care.'

'But the deputy leader of the insurgency was a schoolfriend of yours, wasn't he?' I asked.

'Not exactly a friend,' he said, 'but we were at school together. He was very popular with the Christian missionary teachers. I wasn't.'

'What we call a little swot?' I asked.

'Something like that.'

I asked him what had happened to the men who had kidnapped his daughter.

'She was so brave!' Dev said, almost with tears in his eyes. 'When the kidnappers said they were going to take one of us away, she immediately got up and volunteered. My sixteen-year-old daughter! I felt so helpless. It was really very difficult for me. Why should she be tortured because of my success?'

'What happened?' I asked.

'I had to pay a ransom. But Medha had noticed some details of the place to where she was taken in Patan, because her blindfold had slipped. There was a big police operation and they caught the whole gang. But there was no established sentence for kidnapping – maybe just a year or two in prison. But then the police found drugs on them and they all got fifteen years.'

Dev knew well enough that I was longing to see the high Himalayas, but for much of the time that I was there both the foothills and the mountains beyond them were obscured by mist. We eventually managed to see them in the distance – briefly in the morning before the clouds moved in – from a town called Dhulikhel, after an hour's drive from Kathmandu at the crack of dawn.

The snow-covered mountains seemed to float in the sky, above the mist-hidden foothills and valleys, serene and celestial, and entirely detached from the world below in which I lived. It needed no imagination to think that gods lived there. I wept silently with happiness that I had lived long enough to see them. And then the clouds rose up from the west and in a matter of minutes the mountains had disappeared from view.

On a later trip to Nepal, I took a few days off work to go trekking in the mountains with my son William, who came out to join me for two weeks. We walked for five days, the first day ascending from Nayapul – a typically scruffy, dusty and rubbish-strewn Nepali town – up towards the foothills around Annapurna South, one of the several peaks of the Annapurna range. After a few miles the dust track ends, and from then on you climb up a path paved with rough-hewn stone and what feels like an endless flight of stone steps. We climbed more than 1,000 metres on the first day like this, the temperature in the 80s. William and our guide Shiva – a delightful man, both solicitous and discreet – climbed imperturbably while I streamed with sweat and had to stop at regular intervals to catch my breath. I had assumed that my daily exercise regime meant I was very fit. I am getting old, I thought, and remembered how so many of my elderly patients in England would protest when I explained that their problems were due to old age. 'But Mr Marsh, I still feel so young!'

As we climbed, the villages at first were all formed around small subsistence farms. Shiva would point out the various crops being grown – rice at the lower levels, potatoes and corn as we slowly climbed higher. Annapurna is a conservation area, and entirely free from the rubbish of the towns. There are medieval scenes to be seen – a farmer with a pair of oxen ploughing a sloping field, with steep hills and mountains in the distance, old women carrying firewood in large baskets on their backs, mule trains going up and down the stone steps. As you climb higher the hillsides become too steep and cold for any farming. The entire area is now based on the trekking industry, a very important part of the Nepalese economy. It is a little strange to see the local Nepalis slowly walking up and down the stone stairs, carrying enormous baskets on their backs or tree trunks or building materials, alongside the wealthy Westerners in their shorts and T-shirts and backpacks. I saw a couple of German tourists at one of the guest houses, slowly walking barefoot on the sharp gravel of the footpath outside our guest house. Later I saw them setting off with a group of trekkers all carrying yoga mats, so I suppose they were seeking enlightenment in the high mountains as well as on gravel. A grey-haired English woman, travelling alone, told us she was heading for a remote village.

'It's said that there are old lamas there,' she said with a note of awe in her voice, and then added, 'But they might not talk to me.'

'Twenty per cent houses now empty,' Shiva told us, pointing out yet another empty dwelling. Rural depopulation continues apace, with more and more people leaving for the nearby city of Pokhara. The houses in the mountains are usually built of stone, with wooden balconies, and some still with stone roofs. They can be very beautiful. The roofs are increasingly being replaced with incongruous bright-blue corrugated metal. His own house, Shiva told us, had been badly damaged in the earthquake. He had young children and elderly parents to care for, so he had to spend much of the year guiding trekkers, trying to accumulate enough money to build his family a new home. His life, he said, was rather difficult at the moment, and I thought he looked old and careworn for his age of thirty-three.

We passed many mule trains going up and down the trail,

the patient creatures carrying gas cylinders, concrete blocks, sacks of cement, food and crates of beer. They carefully picked their way over the rough stone steps, their neck bells daintily ringing. We longed to see the high, snow-covered mountains, but they remained stubbornly hidden in cloud and the view was only of the tree-covered foothills – mountains themselves by European standards, and thousands of feet high.

The guest house on the second night was at the trekking village of Ghorepani, at just under 11,000 feet in altitude. It seemed that William and I were the only guests. Our bedroom was like a large tea chest with just enough room for two hard beds, with walls made of plywood and the original manufacturers' stencils in black ink still in place. We spent a friendly evening sitting round a large stove with Shiva and our hosts. The stove – it was quite cold outside by now – was made from an oil drum with a flue going up through the ceiling, with metal bars welded to it for drying clothes. There was a tremendous thunderstorm, the first rain of the season. William and I fell asleep in our tea chest of a room to the sound of the rain beating a stereophonic tin symphony on the roof above our heads. Shiva had expressed the hope that the rain would clear the clouds and we would see the Himalayas in all their glory from nearby Poon Hill at dawn, before the clouds rise from the valleys and hide the mountains. So we had to get up next morning shortly after four.

Although our guest house was empty, many other trekkers – dim and silent figures in the pitch-dark night – suddenly appeared in single file, heading for the hill. We joined their silent procession; it felt more like a dark stampede and was a strangely sinister experience. At the foot of the stone stairs leading to the summit two dogs, snarling and fighting furiously, locked together, came tumbling down the steps in the dark and almost knocked me over. I was gasping for breath in the thin mountain air within minutes and felt as though I was having a panic attack. But I felt compelled upwards by the silent figures in the night around me. They seemed to be climbing the 300 metres of stone stairs to the summit quite effortlessly. All I could hear was my own panting breath and I was soon soaked in sweat. Or perhaps it was my deeply competitive nature that forced me upwards – the thought of being overtaken by anybody being unbearable, even though I was probably the oldest person on the hill. So I hurried, gasping, upwards.

It felt like an ascent to hell, rather than the more conventional descent. We hoped to see the sun rise over the high mountains but they would have none of it and had promptly wrapped themselves in dense cloud. William and I quickly left the crowd on the top of the hill, most of them clutching their smartphones and cameras, hoping to see the mountains. At the height of the trekking season, Shiva told me, there can be many hundreds of people on Poon Hill at dawn. There was some consolation in passing the late arrivals as we descended, and now that we could see them in the daylight, toiling up the stairs, they looked as breathless and tortured as I had felt.

The day was spent walking along a high ridge through a rhododendron forest. The trees were as large as oaks, with mottled and flaking trunks, and must have finished flowering a few days earlier, so we walked over a path of pink and red petals. The next night was spent in a guest house which was supposed to have fine views of the high mountains, but when we arrived there we could only see cloud and foothills. Our bedroom had unglazed windows with elaborately carved, black wooden shutters. I woke in the middle of the night. I could see a few stars through a half-open shutter and so could hope to see the mountains in the morning. I listened to my son's quiet breathing as he slept in the bed next to mine. I thought of his birth thirty-seven years earlier. How he had been placed on his mother's stomach and how he then opened two large, thoughtful blue eyes as he saw the outside world for the first time. I thought of how within a few months he had almost died. Years later we had come close to becoming estranged. He had gone through a difficult time, and I felt paralysed, knowing that I was part of the problem and that the past could not be undone, for all my regrets. His sister Katharine proved to be of much greater help than me. But that terrible time was also now in the past and I soon fell asleep again.

In the morning I woke from my one recurring nightmare – that I am back at university about to take my Finals, after my year of truancy working as a hospital theatre porter, but I have done no work whatsoever. I am filled with dread and panic. I am told anxiety dreams about examinations are quite common, but I find it curious that it is so locked into my subconscious. When I was allowed back to the university after running away, and after my brief stay in the psychiatric hospital – I continued to see my psychiatrist once a week – I worked frenetically hard and got a good degree, so I do not know why this fear of failure so often haunts me when I sleep.

I got out of bed to find that the mountains of the Annapurna range had miraculously appeared, where before there had been only cloud. It's more as though they had suddenly arrived, in complete silence, from somewhere else, that they had descended from heaven. They towered above us, brilliant white with ice falls, snow fields and glaciers, against the blue sky. They looked so close that you felt you could get to them with a walk of only a few hours, when in reality Annapurna base camp, at the foot of Annapurna South, is four days' walk away.

There was then a long walk downhill back to Nayapul, at first along a little-used track in a steep and peaceful wood, with the great mountains still to be seen between the trees, before the clouds rose up from the valleys and the mountains disappeared. We had to stop at regular intervals to kick the leeches off our boots. Later we rejoined the stone path and steps, passing many mule trains climbing up in the opposite direction. As we slowly descended we were accompanied by the sound of the glacial grey and white River Modi Kholi, rushing over rocks far below us.

'Twenty-two-year-old fell thirty metres. Caesarean section. On examination no movement in lower limbs and weak upper limbs.' The MO rattled and stumbled through the presentation.

'Oh, come on!' I shouted. 'That's a hopeless presentation. How much movement does she have in her arms? What's her spinal level?'

We worked out that she had only partial movement in her biceps and none in her triceps, that she could weakly shrug her shoulders and bend her elbows, but that everything below this – her hands, her spinal and abdominal muscles and her legs, her bowels and bladder – was all completely paralysed.

'So her spinal level is C5/C6. Yes? And don't you have any curiosity as to what's happened to your patients? Thirty metres? How can one survive that? Was it suicide? And was this after the caesarean section?'

'She fell off cliff while cutting grass with sickle. Foetal death, so caesarean section. Then she came to Neuro Hospital.'

'How many months pregnant?'

'Seven months. Husband working in Korea.'

'Ah,' I said, appalled. 'Well, let's look at the scan.'

The MRI scan showed fracture and complete translocation of the spinal column between the fifth and sixth cervical vertebrae. The spinal cord looked damaged beyond repair.

'She'll never recover from that,' I said. 'What's the next case?'

Dev and one of the registrars operated the next day, screwing the girl's broken spine back together again, although this could not undo the paralysis. Surgery would at least mean that she did not need to be kept flat on her back in one of those horrible cervical collars, and it would make the nursing and physio easier.

I saw her on the ITU next morning as Dev and I went round.

'Presumably here in Nepal she'll get bedsores and renal infections if she ever gets out of hospital?' I said.

Dev grimaced.

'She's unlikely to survive long. Christopher Reeve was a millionaire and lived in America and he eventually died from complications, so what chance a poor peasant in Nepal?'

I looked at the girl as we talked – at least she couldn't understand what we were saying. She was very beautiful in the way that so many Nepali women are, with large, dark eyes and high cheekbones and a perfectly symmetrical and outwardly serene face. Her eyes moved slowly, she spoke a few words when spoken to. Her head was immobilized in a large and uncomfortable pink plastic surgical collar. Dev agreed with my suggestion that it could be taken off now that she had her broken neck screwed and plated back together again.

'I put a locking plate in,' Dev said. 'Very expensive. Thousands of rupees.' He then launched into a tirade once again about the way the foreign equipment companies charged First World prices in Third World countries and how most surgeons using implants would be paid a 20 per cent kickback by the suppliers, the extra cost being passed on to the patient. He said he had always refused to get involved in this widespread, but thoroughly corrupt, practice. You can find it in many European countries as well, despite being illegal, although there the inflated extra cost can often be passed on to the taxpayer and government rather than to the patient.

'Well, medical-equipment manufacturers are businessmen, not altruists,' was all I could say.

After a few days on the ITU the paralysed girl was discharged to one of the wards, but shortly afterwards her breathing deteriorated – which often happens in these cases – and she had to be readmitted and put on a ventilator.

'I spoke to her husband again yesterday,' Dev told me. 'He's flown back from Korea. I think he is coming to accept that she might die. But it's very difficult in Nepal. If you are too honest and realistic it causes terrible trouble. The family will be shouting and screaming all over the hospital and causing all sorts of problems. You just can't tell them the truth straight out. I told him he was young. I said that if she dies he could at least start again.'

'It's easier now that she's on a ventilator, isn't it?' I replied, because it would be kinder if she died anaesthetized on a ventilator than from bed sores and infection on a bed in the hospital or back in her home – not that she was likely ever to get home.

On the morning round next day I noticed a group of doctors and nurses round the girl's bed. She was groaning terribly as an anaesthetist pushed a flexible, fibre-optic bronchoscope down her trache tube. Her chest X-ray looked awful. We watched the intriguing view of the ringed and ridged inside of her lungs' bronchi on the small monitor attached to the bronchoscope, while she groaned piteously as the anaesthetist tried to clear the fluid from her lungs. We agreed she was better off dying, but Dev was in an impossible situation. Should he have refused to operate and left the woman with her dislocated, broken neck untreated, leaving her to die without any treatment? The family would almost certainly have refused to accept this. Should he have left them to take her to another hospital where she would have undergone surgery that probably would not have been done as well as it would have been in his hospital? I had never had to face problems like this in my own career.

We get so used to most of our patients having brain damage and being unconscious that we forget that some of the paralysed patients on ITUs are wide awake, suffering horribly but unable to show it. Or perhaps it is wilful blindness on our part. I was painfully aware that I had found some of these cases so distressing during my career that I tended to avoid them and walk past them on the ward round. What do you say to somebody who is completely paralysed from the neck down, but awake, on a ventilator, so that they cannot talk?

I remembered an identical case in Ukraine many years ago. My colleague Igor was still working in the government emergency hospital at the time. He was very proud of the fact that he had managed to keep the patient alive, but on a ventilator.

'First case of long-term ventilation in Ukraine,' he declared.

The young man was in a bleak little side room and lived there for three years. Many religious icons surrounded him on the otherwise bare walls. He was equipped with a speaking tracheostomy tube and each time I visited Igor's department I would go and see him. His brother looked after him and spoke some English, so I communicated with the patient through him. Each time I saw him he had wasted away a little more. At the time of the injury – breaking his neck diving into shallow water – he had been quite heavily built, but by the time he died he was skin and bones. At first

I was able to have quite rational conversations with him, but it became more difficult with each visit. At least, he started to ask me about religious miracles and salvation, which he spoke about with intense passion (to the extent that you can speak passionately with a speaking tracheostomy tube), to which I had no answer. I was relieved on a later visit to see that the little side room was empty.

The young Nepali woman had fallen and broken her neck during *Dasain*, the most important of the many Nepali festivals, when upwards of fifty thousand goats and hundreds of buffalo are sacrificed to the goddess Durga. Blood is smeared everywhere in honour of the goddess, including, I noticed, on Dev's gold-coloured Land Rover. Animal rights activists, I read in a local newspaper, have recently suggested that the goats be replaced with pumpkins.

The festival goes on for two weeks. Two days earlier Dev had told me to accompany him to the gates in front of his house. A police jeep was parked there with a uniformed policeman standing beside it. Another policeman appeared, leading a beautiful goat with long, floppy ears on a rope from behind the garage.

'I give the local police a goat every year for *Dasain*,' Dev told me. The goat was bundled into the back of the jeep but immediately jumped out. So it was put back in, but now with a police escort. They drove away with the goat looking mournfully out at me over the tailgate, the policeman beside it.

'That goat will feed a hundred policemen,' Dev said approvingly.

'Nobody is in the mood for *Dasain*, this year, what with the earthquake and now the blockade and fuel crisis,' Dev commented as we drove back to Kathmandu from a visit to a nearby town. Yet in several places we passed the beautiful high swings – known as *pings* – which are a traditional part of *Dasain* celebrations. They are made simply of four bamboo poles lashed together, more than twenty feet high and decorated with colourful flags. I saw Nepalis – both adults and children – laughing ecstatically as they swung happily to great heights, although I thought the *pings* looked a little precarious.

The next day I sat in the library teaching the juniors and discussing how we could improve the MOs' jobs.

'I am going back to London tomorrow,' I told the new cohort of MOs, freshly out of medical school and, it seemed to me, pretty well out to lunch.

'You are good doctors. We want to make you better. I hope the registrars' – I looked pointedly at them – 'will try to continue the morning meetings in this spirit. Teasing, yes, but no bullying.' Pleased with this little speech, I then went down to Dev's office and was about to go downstairs to start the clinic when there was a sudden flurry of activity in the corridor outside.

I found Dev, looking grim, surrounded by several of his juniors at the theatre reception desk, all looking equally serious.

'The girl with a broken neck has just died,' Protyush told me. 'The husband is very angry.'

'Is Dev waiting to talk to him?'

'Yes, but we need backup – here in Nepal the families can assault us. We're waiting for the security guards.'

Thirty minutes later, I stood in a corner of the theatre reception area where I had a view into the counselling room, and I could see Dev, but not the angry husband. Dev listened to a long outburst in silence and spoke quietly in reply. I crept away, not liking to eavesdrop on so much tragedy and unhappiness.

'I wish I still worked for the NHS,' Dev said to me that

evening, as we sat in the garden. 'Or at least that I was still the only neurosurgeon here, or that I didn't have to worry about keeping the hospital afloat financially. It's yet to make a profit, you know, even after ten years. Twenty years ago I could simply have said that there was nothing to be done and the family would have accepted it.'

'How did the meeting with the family go?' I asked.

'Oh – the usual stuff. It happens now every few months. Never happened in the past. The husband said I had killed his wife by doing a tracheostomy. Nonsense of course – and in fact, in six months' time, he'll probably have a new wife. If she had survived it would have been terrible for both of them. And I spent so long, every morning, trying to explain. And he was so polite, as though I was a god, but now I'm a devil. But I'm sure you'll find there's another neurosurgeon in town who's told them that if he had treated her she'd have been OK.'

'You can't expect people to be reasonable immediately after a death like this one,' I said, trying to be helpful.

'Nepal is different,' Dev replied. 'I worry for the boys, when they become seniors, having to work in a country like ours where the people are so uneducated – they won't have my authority. All the hospitals have a permanent plain-clothes policeman stationed twenty-four hours a day because of problems like this. They said they would get all the other patients' families to blockade the hospital. Said they would burn it down. They want money. I know a lot of other doctors here who have had money extorted from them. That's the problem with having to run a private hospital – "We paid you to treat her," they said, "and now she's dead." It was so much easier in the past when I worked at the Bir. But the government medical service here now is terrible, almost completely broke. And so when I first see a patient the initial question is not what treatment would be best for them but "What can you afford?" You're so lucky to work in the NHS.'

'Well, she's better off dead,' I said.

It was sad to see Dev – usually so cheerful and enthusiastic – suddenly silent, looking grim.

'You can't really share it with anyone. It would only upset and frighten my wife,' he added.

'Only neurosurgeons understand,' I said, 'how difficult it is to be so hated, especially when you haven't even done anything wrong, and only tried to do your best.'

I remembered one of my first catastrophes as a consultant. A child who died as a result of my postponing an operation that should have been done urgently. I had thought it was safe to wait until the morning, but I had been wrong. I had to attend an external investigation. I did not have to meet the parents face to face but passed them in the corridor. The look of silent hatred the mother gave me was not easy to forget.

'You start,' he said, pointing to the bottle of beer I had already got out. 'The woman's MP might come round to the hospital – I don't want to smell of alcohol.'

I was summoned to supper two hours later. To my surprise, all the managerial team of the hospital were present – six people including the driver, all there to support Dev. I was rather touched. I'd never had support like this for my disasters.

Over a large Nepali dinner there was much animated discussion, most of it lost on me as they spoke in Nepali. But I was told that the family were threatening a hunger strike and a press conference, and planned to get the other patients' families to join them.

'Seven point five,' I heard the manager, Pratap, suddenly

say – he had been looking at his smartphone. This, it turned out, was the strength of an earthquake that had just hit Afghanistan and Pakistan. The catastrophic earthquake that had hit Nepal six months before my visit had been 7.8. This was discussed for a while, and then they resumed the conversation about the dead girl's family and what might happen.

'It's all because we now work for money,' Madhu, who was sitting next to me, said. 'We didn't want to, but had no choice. We can't provide free treatment to everybody.'

Next morning, the morning of my departure from Nepal, I sat drinking coffee in the garden, in Dev and Madhu's little Shangri-La. The pigeons were cooing and gurgling, the cocks were crowing, the hooded crows were quarrelling again in the camphor tree, although in truth for all I knew they might have been discussing their marital problems or the presence of the brown mongoose which can sometimes be seen, sinuous and graceful, running swiftly across the garden. Or perhaps they were excited about the prospect of the first day of the festival of *Tihar* in two weeks' time, the day of *kaag tihar*, when crows are worshipped and little dishes of food are put out for them. I probably understand as much about the crows as I do about the impenetratable complexities of Nepali society. Two birds with feathery trousers I couldn't identify waddled busily about on the small lawn in front of the gazebo.

I set off for work as usual but as it was the tenth and most auspicious day of *Dasain,* there was little traffic on the road. I passed women wearing their finest clothes – brilliants reds and blues and greens, decorated with gold and silver and paste jewels which flashed in the sunlight. They picked their way cautiously over the puddles and around the rubbish and stinking, open drains. When I got to work I found that there were twelve uniformed policemen with long iron-shod sticks in front of the hospital, sitting in the sunshine on the grass mound by the magnolia tree. The dead woman's family and supporters stood nearby. Dev and I looked down at them from his office window.

'How much longer will this go on for?' I asked.

'Oh, until the weather gets colder,' he said with a laugh, his cheerful good humour having returned.

'I'm not even sure the story about cutting grass on a cliff was true. Her husband has money – it's unlikely she'd be out gathering grass off a cliff,' he said. 'I'm pretty suspicious that it was another *ping* accident.'

We had admitted a sixty-five-year-old man two days earlier, also completely paralysed, with a broken neck, who had fallen from a *ping*.

'Happens all the time during *Dasain*,' Dev said.

I noticed that behind the policemen, the waiting outpatients and the dead woman's angry family, in the rice paddy next to the hospital, people were harvesting the rice – a picturesque and medieval sight, although in the background there was a long queue of dirty old trucks waiting at the petrol station. In the distance, the high Himalayas, beyond the foothills, were hidden.

8

LAWYERS

I had to return to London from Nepal earlier than I had originally planned because I was due to appear in court. A patient was suing me. The case had been dragging on for four years. I had operated for a complex spinal condition causing progressive paralysis, and the patient had been initially left worse than he had been before the operation. As far as I could tell he had eventually ended up better than before the operation, but apparently he was deeply aggrieved. A neurosurgeon – justly famous for the very high opinion he had of himself, although less famous for his medico-legal pronouncements – was of the opinion that I had acted negligently. Just for once, I was as certain as I could be that I had not, and I had reluctantly felt obliged to defend myself. It was just like Nepal, I thought. All these surgeons attacking each other. I had had to attend various meetings about the case and many thousands, probably hundreds of thousands, of pounds must have been spent in legal fees. At the last moment, after I had come all the way back from Nepal, the claimant and his lawyers abandoned the case two days before the trial was due to start. The solicitor handling my defence was most apologetic about the waste of my time.

'But it's better than needing twelve policemen,' I replied cheerfully, without explaining what I meant.

Many doctors do what is called medico-legal work, providing reports for lawyers in cases involving personal injury or medical negligence. It is a lucrative but time-consuming business. I did a few such reports myself when I became a consultant, but quickly gave it up. I preferred operating and dealing with patients to the many meetings and lengthy paperwork which medico-legal work requires. I only became involved with lawyers if I was being sued myself – always a very distressing experience, whether I felt guilty or not.

This occurred four times during my career, including the case which had forced me to return from Nepal and which had now collapsed. The other three cases had all been settled, as I blamed myself for what had happened and did not want to defend myself. One case had been for a retained swab after a spinal operation (in the days when swab counts were not being done in the old hospital) which had not caused any severe injury, and the other two were cases where I had been slow to diagnose serious, although almost uniquely rare, post-operative infections. One of those patients had come to serious harm, the other to catastrophic harm.

But a few years ago I was subpoenaed to give evidence in a personal injury compensation case, which I regarded as an absurd and complete waste of my time. So I attended reluctantly, a series of High Court orders having been served on me over the three days before the hearing. The men serving them had never been able to serve them on me in person – which, strictly speaking, I believe the law requires. The first attempt had been made while I was operating and the second when I was away from London the following day. I returned the following evening to find that a copy of the order had been pushed through the letterbox of the front door of my home. I was operating the day after that until the evening, and came out of the theatre to be told that earlier in the morning a man had walked up to the hospital reception desk and had thrown down yet another copy of the High Court order in front of the receptionist and then stalked off.

This barrage of court orders had been unleashed upon me by a solicitor in a huge City law firm which was acting on behalf of an American law firm, which in turn was acting for the defendants in the compensation case.

An English woman had been involved in a minor car accident in the USA while on holiday, and had subsequently seen me as a patient about her 'whiplash' symptoms. I had confirmed with an MRI scan that there were no significant injuries to her neck and reassured her that she would get better in time. In practice it is not at all clear whether these whiplash syndromes do get better. Patients develop an array of aches and pains and altered sensations in their necks and arms which do not correspond to any known pathological processes such as bone fractures or torn muscles or trapped nerves, and do not spontaneously improve in the time it takes most proven 'soft-tissue' injuries to heal and become pain-free. It is well known that these syndromes do not occur in countries which do not have any legal recognition of whiplash injury as a consequence of minor car crashes.

The particular type of accident which is alleged to produce 'whiplash injury' is a 'shunt', when a car is driven into from behind by another car. These are typically low-speed accidents, where the driver or passengers are subjected to relatively slight forces, never enough to cause any obvious injuries, but which seem to produce severe and lasting symptoms without any evidence of injury such as bruising or swelling or changes on an X-ray or MRI scan. It has been pointed out that driving dodgem cars on fairgrounds involves near-continuous shunting as the cars are deliberately driven into each other, and yet there are no reports of whiplash symptoms afterwards. This discrepancy between the severity of the symptoms and the apparent triviality of the injury has been attributed to a putative 'whiplash' effect. The victim's neck is supposed to be cracked like a whip – something that has never in fact been demonstrated and is probably fallacious.

I used to see many of these patients every year in my out-patient clinic and it was clear to me that most of them were not consciously malingering – instead they were the willing, perhaps hapless, victims of a 'nocebo' effect, the opposite of the placebo effect. With the placebo effect, which is well understood, people will feel better, or suffer less pain, simply as a result of suggestion and expectation. With 'whiplash injury', the possibility of financial compensation for the victims, combined with the powerful suggestion that they have suffered a significant injury, can result in real and severe disability, even though it is, in a sense, purely imaginary. They are more the victims of the medico-legal industry and of the dualism that sees mind and brain as separate entities than of any physical injury outside the brain. It is the modern equivalent of the well-attested phenomenon of a witch doctor in tribal society casting a spell on somebody, causing the victim to fall ill, merely through the power of suggestion and belief. There was a further significant irony in this case, which I had mentioned in my original letter about the patient: the victim's husband was a lawyer specializing in personal injury compensation.

I had been given only two weeks' notice about the hearing – strictly speaking, the 'deposition of evidence before a Court-appointed Examiner'. I was told that I was required to attend but there was no mention of legal compulsion. My secretary had told the woman solicitor who had sent the letter that I could not attend as I was already committed to operations and outpatient clinics. As I had heard nothing more after my secretary had told the solicitor this, I had assumed that it had been accepted that I would not be coming. It seems that the solicitor, however, decided that I needed to be taught a lesson and served me with the court order. I had some urgent cases to do, which could not be postponed. I therefore started operating at seven in the morning on the day of the deposition, at high speed, something I hate doing; nor had I slept well, as I was angry that I was being dragged away from my work in this way.

I was not going to be paid, but doubtless the lawyers would be paid hundreds of pounds, probably thousands, for trying to extract a medical opinion from me for free. I knew the business would be absurd – I had seen the patient only twice, four years ago, had no memory of her whatsoever, and the lawyers already had copies of my correspondence. I clearly would have nothing to add. So I was angry, and had already telephoned the solicitor the day before and told her so.

The law firm's offices were housed in a huge postmodern marble and glass office block just beyond the Tower of London. I marched into the building full of righteous indignation, past the men in suits smoking cigarettes on the piazza outside, and clutching my folding bike and attaché case. I collected a laminated visitor's pass from a receptionist in a smart uniform, pushed past the barricade of the revolving stainless-steel turnstile and ascended to the seventh floor in one of the many tall, swift lifts lined with dark mirrors. If only my hospital had such lifts – how much time it would save!

I emerged into a three-storey-high atrium, walled and

floored in marble, even though already on the seventh floor. High plate-glass windows showed a panoramic view over the City towards the Lloyd's Building and the various high and imposing office blocks around it. Having announced myself, I had to wait for a while, and looked with sour awe at the City under a clear blue sky. Babylon! I thought – the heart of an extravagant culture, consuming itself and the planet, sheathed in glittering glass. A slim and polished barrister in a light-charcoal pinstriped suit, the Court-appointed Examiner, descended the elaborate glass, steel and hardwood spiral staircase at one side of the atrium and introduced himself. He was, perhaps, just a little apologetic and thanked me for coming.

'I am not pleased to be here,' I growled.

'Yes, so I heard,' he answered politely.

He led me to an anonymous, luxurious and windowless meeting room, the furniture all in white ash and chrome, where the English QC for the plaintiff and the American lawyer for the defendants were waiting for me. The American lawyer was in his fifties and was fit and trim, with short grey hair and a designer sports jacket. The elderly English QC, however, did not look as though he worked out in a gym every day and was rather overweight, with a florid face, and wore a crumpled white linen suit and half-moon glasses.

'Good morning gentlemen,' I said as I entered, feeling a little superior, knowing that they were not going to get anything out of me. I sat down and after the introductions a man with a video camera read out, in a bored voice, the description of the proceedings. I was sworn in (I affirmed rather than swore on the tatty little Bible on offer) and briefly cross-examined. This could only mean that I could agree that the notes I made four years ago were indeed mine and that I had no memory of the case. The American lawyer, of course, wanted to extract my opinion about whiplash injury, but I refused to be drawn.

'It is a medico-legal question,' I said, 'and I therefore have no opinion. I never give medico-legal opinions over personal injuries.' Whether they heard the disdain in my voice or not, I do not know.

I had seen the patient and had advised against surgery. The English QC wanted me to agree that if her symptoms had not got better as I had said they probably would, it was reasonable for her to seek a further opinion. I agreed that it was.

'Did you know,' the American lawyer then asked, 'that she did eventually undergo surgery on her neck?'

'No,' I said.

How much I could have said! I had affirmed that I would tell the truth, the whole truth and nothing but the truth, but not that I would not be economical with it. I could have explained the psychosomatic nature of whiplash injury, the nonsense written about the alleged mechanism, the fact that all the neurosurgical textbooks state that one should never operate on the spine of somebody involved in compensation litigation. They never, ever get better. Some greedy surgeon must have operated on her neck and now, most probably, her symptoms were even worse and the lawyers would be arguing over whether her disability was the result of the original trivial injury or the operation. I could have told the lawyers that they themselves were more responsible for her problems than the original minor car crash. The principal consequence of that trivial accident, and the millions of other ones like it, was not just the plaintiff's pain and suffering, but also the Babylonian marble offices where we were now meeting. The humourless men seated round the table before me were part of the great industry of personal injury compensation, with its army of suave and accomplished lawyers and assured expert witnesses, rooting in a great trough of insurance premiums.

At the end of the meeting the American lawyer went through my CV, which he had in his hand. His face was impassive but he seemed a little puzzled by it. I am rather proud of my CV and academic record, and I thought that perhaps he too would be impressed by it and would be arguing that since an English surgeon with such a brilliant CV had advised against surgery, the operation carried out by somebody else could not have been a good idea.

'How did you get all those prizes at college?' he eventually asked.

'By working very hard,' I replied, feeling deflated. He remained expressionless – perhaps he was just bored and wanted a little distraction – but the English QC smiled.

And that was that. The video camera was switched off and the Examiner thanked me for coming.

'Well, I'll get on with my day,' I said. I descended the spiral staircase, collected my folding bike from the reception desk and left.

9

MAKING THINGS

A long time ago I had promised my daughter Sarah that I would make her a table. I am rather good at saying I'll make things, and then finding I haven't got the time, let alone getting round to make the many things I want to make or mend myself.

A retired colleague, a patient of mine as well, whose back I had once operated upon, had come to see me a year before I retired with pain down his arm. Another colleague had frightened him by saying it might be angina from heart disease – the pain of angina can occasionally radiate down the left arm. I rediagnosed it as simply pain from a trapped nerve in his neck that didn't need treating. It turned out that in retirement he was running his own oak mill, near Godalming, and we quickly fell into an enthusiastic conversation about wood. He suggested I visit, which I did, once I had retired. To my amazement I found that he had a fully equipped industrial sawmill behind his home. There was a stack of dozens of great oak trunks, twenty foot high, beside the mill. Eighty thousand pounds' worth, he told me when I asked. The mill itself had a fifteen-foot-long sawbed on which to put the trunks, with hydraulic jacks to align them, and a great motorized bandsaw that travelled along the bed. The tree trunks – each weighing many tons – were jostled into place using a specialized tractor. All this he did by himself, although in his seventies, and with recurrent back trouble. I was impressed.

I spent a happy day with him, helping him to trim a massive oak trunk so that it ended up with a neatly square cross-section, and then rip-sawing it into a series of thick two-inch boards. The machinery was deafening (we wore ear defenders), but the smell of freshly cut oak was intoxicating. I drove home that evening like a hunter returning from the chase, with the planks lashed to the roof rack of my ancient Saab – a wonderful car, the marque now, alas, extinct – that has travelled over 200,000 miles and only broken down twice. The roof rack was sagging under the weight of the oak and I drove rather slowly up the A3 back to London.

The next morning I went to collect my bicycle from the bicycle shop in Wimbledon Village, as it likes to be called, at the top of Wimbledon Hill. Brian, the mechanic there, has been looking after my bicycles for almost thirty years.

'I'm afraid the business is closing down,' Brian told me, after I had paid him.

'I suppose you can't afford the rates?'

'Yes, it's just impossible.'

'How long have you been here?'

'Forty years.'

He asked me for a reference, which I said I would gladly give. He is by far the best and most knowledgeable bike mechanic I have ever met.

'Have you got another job?' I asked.

'Delivery van driver,' he replied with a grimace. 'I'm gutted, completely gutted.'

'I remember the village when it still had real shops. Yours

is the last one to go,' I said. 'Now it's all just wine bars and fashion boutiques. Have you seen the old hospital just down the road where I worked? Nothing but rich-trash apartments. Gardens all built over, the place was just too nice to be a hospital.'

We shook hands and I found myself hugging him, not something I am prone to do. Two old men consoling each other, I thought, as I bicycled down the hill to my home. Twenty years ago I lived with my family in a house halfway up the hill. I assume that the only people who can afford to live in the huge Victorian and Edwardian villas at the top of the hill are bankers and perhaps a few lawyers. After divorce, of course, surgeons move to the bottom of the hill, where I now live when not in Oxford or abroad.

The oak boards needed to be dried at room temperature for six months before I could start working on them, so I clamped them together with straps to stop them twisting and left them in the garage at the side of my house (yet another of my handmade constructions with a leaking roof), and later brought them into the house for further drying.

Now that I was retired and back from Nepal, the wood was sufficiently dry for me to start work.

When my first marriage had fallen horribly apart almost twenty years earlier and I left the family home, I took out a large mortgage and bought a small and typical nineteenth-century semidetached house, two up and two down, with a back extension, at the bottom of Wimbledon Hill.

The house had been owned by an Irish builder, and his widow sold the house to me after his death. I had got to hear that the house was for sale from the widow's neighbours, who were very good friends of mine. So the house came with the best neighbours you could wish for, a wide and unkempt garden and a large garage in the garden itself, approached by a passage at the side of the house. Over the next eighteen years I subjected the property to an intensive programme of home improvements, turning the garage into a guest house (of sorts) with a subterranean bathroom, and building a workshop at the end of the garden and a loft conversion. I did much, but not all, of the work myself. The subterranean bathroom seemed a good idea at the time, but it floods to a foot deep from an underground stream if the groundwater pump I had to have installed beneath it fails.

The loft conversion involved putting in two large steel beams to support the roof and replacing the existing braced purlins (I had taken some informal advice from a structural engineer as to the size of steel beam required). With my son William's help I dragged the heavy beams up through the house and, using car jacks and sash cramps, manoeuvred them into position between the brick gables at either end of the loft. There was then an exciting moment when, with a sledgehammer, I knocked out the diagonal braces that supported the original purlins. I could hear the whole roof shift a few millimetres as it settled onto the steel beams. I was rather pleased a few years later to see a loft conversion being done in a neighbouring house – a huge crane, parked in the street, was lowering the steel beams into the roof from above. I suppose it was a little crazy of me to do all this myself, and I am slightly amazed that I managed to do it, although I had carefully studied many books in advance. The attic room, I might add, is much admired and I have preserved the chimney and the sloping roof, so it feels like a proper attic room. Most loft conversions I have seen in the neighbourhood just take the form of an ugly, pillbox dormer.

I have always been impatient of rules and regulations and sought neither planning nor building regulation permission for the conversion, something I should have done. This was to cause problems for me when I fell in love with the lock-keeper's cottage. I could only afford to buy it if I raised a mortgage on my house in London (I had been able to pay off the initial mortgage a few years earlier). The London house was surveyed and the report deemed it fit for a mortgage, 'subject to the necessary permits' for the loft extension from the local council, which, of course, I did not have.

With deep reluctance I arranged for the local building inspectors to visit. I expected a couple of fascist bureaucrats in jackboots, but they couldn't have been nicer. They were most helpful. They advised me how to change the loft conversion so as to make it compliant with the building regulations. The only problem was that the property developers who were selling the lock-keeper's cottage were getting impatient. So, over the course of three weeks, working mainly at night as I had not yet retired, I removed a wall and built a new one with the required fire-proof door, and installed banisters and handrails on the oak stairs – the stairs on which I had once slipped and broken my leg. I also installed a wirelessly linked mains-wired fire alarm system throughout the house. This last job was especially difficult as over the years I had laid oak floorboards over most of the original ones. Running new cables above the ceilings for the smoke alarms involved cutting many holes in the ceilings and then replastering them. But after three weeks of furious activity it was all done, and I am now the proud possessor of a 'Regularisation Certificate' for the loft conversion of my London home, and I also own the lock-keeper's cottage.

As soon as I had moved into my new home in London seventeen years ago, after the end of my first marriage, I had set about building myself a workshop at the bottom of the garden, which backs onto a small park and is unusually quiet for a London home. I was over-ambitious and made the roof with slates and, despite many efforts on my part, I have never been able to stop the roof leaking. I cannot face rebuilding the whole roof, so two plastic trays collect the water when it rains, and serve as a reminder of my incompetence. Here I store all my many tools, and it was here that I started work on Sarah's table. In the garden, which I have allowed to become a little wild, I keep my three beehives. London honey is exceptionally fine – there are so many gardens and such a variety of flowers in them. In the countryside, industrial agriculture and the use of chemical fertilizers, pesticides and herbicides have decimated the population of bees, as well as the wild flowers on which they once flourished.

It took many weeks to finish the table, sanded a little obsessionally to 400-grit, not quite a mirror finish, using only tung oil and beeswax to seal it. The critical skill in making tabletops is that the edges of the boards should be planed so flat – I do it all by hand – and the grain of the wood so carefully matched that the joints are invisible. You rest the planed edges of the boards on top of each other with a bright light behind them so that a gap of even fractions of a millimetre will show up. This requires a well-sharpened plane. A well-sharpened and adjusted plane – 'fettled' is the woodworker's traditional word for this – will almost sing as it works and minimal effort is required to push it along the wood.

It took me a long time to learn how to sharpen a plane properly. It now seems obvious and easy and I cannot understand why I found it difficult in the past. It is the same when I watch the most junior doctors struggling to do the simplest operating, such as stitching a wound closed. I cannot understand why they seem to find it so difficult – I become impatient. I start to think they are incompetent. But it is very easy to underestimate the importance of endless practice with practical skills. You learn them by doing, much more than by knowing. It becomes what psychologists call *implicit memory*. When we learn a new skill the brain has to work hard – it is a consciously directed process requiring frequent repetition and the expenditure of energy. But once it is learnt, the skill – the motor and sensory coordination of muscles by the brain – becomes unconscious, fast and efficient. Only a small area of the brain is activated when the skill is exercised, although at the same time it has been shown, for instance, that professional pianists' brains develop larger hand areas than the brains of amateur pianists. To learn is to restructure your brain. It is a simple truth that has been lost sight of with the short working hours that trainee surgeons now put in, at least in Europe.

The boards are glued together using what is called a rubbed joint – the edges rubbed against each other to spread the glue – and then clamped together for twenty-four hours with sash cramps. The frame and legs are held fast with pegs, and being oak, the table is very solid and heavy. I had taken great care, when sawing the wood with my friend, that it was 'sawn on the quarter', so that the grain would show the beautiful white flecks typical of the best oak furniture. Sarah was very happy with the result after I delivered it, and subsequently sent me a photograph of her eighteen-month-old daughter Iris sitting up to it, smiling happily at the camera as she painted pictures with paintbrush and paper. But, just like surgery, there can be complications, and to my deep chagrin a crack has recently developed between two of the jointed planks of the tabletop. I cannot have dried the wood sufficiently, I was impatient yet again. I will, however, be able to repair this with an 'eke' – a strip of wood filling the crack. It should be possible to make it invisible, but I will probably have to refinish the whole surface.

*

I'm not sure how my love of and obsession with making things arose. I hated woodwork at school: you had no choice as to what you made and you would come home at the end of term with some poorly fashioned identikit present for your parents – a wobbly little bookcase, a ridiculous egg-rack or a pair of bookends. I found these embarrassing; my father was a great collector of pictures, antiques and books and there were many fine things in the family home, so I knew how pathetic were my school woodwork efforts. He was also an enthusiastic bodger who loved to repair things, usually involving large quantities of glue, messily applied. The family made ruthless fun of his attempts, but there was a certain nobility to his enthusiasm, to his frequent failures and occasional successes.

He was a pioneer of DIY before the DIY superstores came into existence. I once found him repairing the rusted body-work of his Ford Zephyr by filling the holes with Polyfilla, gluing kitchen foil over the filler, and then painting it with gloss paint from Woolworth's. It all fell off as soon he drove the car out of the garage. My first attempts at woodwork away from school were made using driftwood from the beach at Scheveningen in Holland, where we lived when I was between the ages of six and eight. I sawed the wood, bleached white by the sea, into the shapes of boats. I made railings from small nails bought at the local hardware store. The only Dutch words I ever learnt were '*kleine spijkes, alsjeblieft*' – small nails, please. I would take these boats sailing with me in the bath, but they invariably capsized.

When I married my first wife, we had no furniture and little money. I made a coffee table from an old packing case with a hammer and nails. It was a wooden one from Germany, with some rather attractive stencilled stamps on it, a little reminiscent of some of Kurt Schwitters' *Merz* work. It had been sitting in my parents' garage for years and had contained some of the last possessions of my uncle, the wartime Luftwaffe fighter pilot and wonderful uncle who eventually died from alcoholism many years after the war.

My brother admired the coffee table and asked me to make one for him, and I said I would, for the price of a plane, which I could then buy and use to smooth the wood. I have not looked back since. My workshop is now stacked with tools of every description – for woodwork, for metalwork, for stone-carving, for plumbing and building. There are three lathes, a radial arm saw, a bandsaw, a spindle moulder and several other machine tools in addition to all the hand tools and power tools. I have specialist German bow saws and immensely expensive Japanese chisels, which are diabolically difficult to sharpen properly. One of my disappointments in life is that I have now run out of tools to buy – I have acquired so many over the years. Reading tool catalogues, looking for new tools to buy – 'tool porn', as my anthropologist wife Kate calls it – has become one of the lost pleasures of youth. Now all I can do is polish and sharpen the tools I already have, but I would hate to be young again and have to suffer all the anxieties and awkwardness that came with it. I have rarely made anything with which I was afterwards satisfied – all I can see are the many faults – but this means, of course, that I can hope to do better in future.

I once made an oak chest with which I was quite pleased. I cut the through dovetail joints at the corners by hand, where they could be seen as proof of my craftsmanship. The best and most difficult dovetail joints, on the other hand – known as secret mitre dovetail joints – cannot be seen. True craftsmanship, like surgery, does not need to advertise itself. A good surgeon, a senior anaesthetist once told me, makes operating look easy.

*

When I see the tidy simplicity of the lives of the people living in the boats moored along the canal by the lock-keeper's cottage, or the sparse homes of the Nepalese peasants William and I walked past on our trek, I cannot help but think about the vast amount of clutter and possessions in my life. It is not just all the tools and books, rugs and pictures, but the computers, cameras, mobile phones, clothes, CDs and hi-fi equipment, and many other things for which I have little use.

I think of the schizophrenic men in the mental hospital where I worked many years ago. I was first sent to the so-called Rehabilitation Ward, where attempts were being made to prepare chronic schizophrenics who had been in the hospital for decades for life in community care outside the hospital. Some of them had become so institutionalized that they had to be taught how to use a knife and fork. My first sight of the ward was of a large room with about forty men, dressed in shabby old suits, restlessly walking in complete and eerie silence, in circles, without stopping, for hours on end. It was like a march of the dead. The only sound was of shuffling feet, although occasionally there might be a shout when somebody argued with the voices in his head. Many of them displayed the strange writhing movements called 'tardive dyskinesias' – a side effect of the antipsychotic drugs that almost all of them were on. Those who had been treated with high doses of a drug called haloperidol – there had once been a fashion for high-dose treatment until the side effects became clear – suffered from constant and grotesque movements of the face and tongue. Over the next few weeks, before I was sent to work on the psychogeriatric ward, I slowly got to know some of them as individuals. I noticed how they would collect and treasure pebbles and twigs from the bleak hospital garden and keep them in their pockets.

They had no other possessions. Psychologists talk of the 'endowment effect' – that we are more concerned about losing things than gaining them. Once we own something, we are averse to losing it, even if we are offered something of greater value in exchange. The pebbles in the madmen's pockets became more valuable than all the other pebbles in the hospital gardens simply by virtue of being owned.

It reminds me of the way that I have surrounded myself with books and pictures in my home, rarely look at them, but would certainly notice their absence. These poor madmen had lost everything – their families, their homes, their possessions, any kind of social life, perhaps their very sense of self. It often seems to me that happiness and possessions are like vitamins and health. Severe lack of vitamins makes us ill, but extra vitamins do not make us healthier. Most of us – I certainly am, as was my father – are driven to collect things, but more possessions do not make us happier. It is a human urge that is rapidly degrading the planet: as the forests are felled, the landfill sites grow bigger and bigger and the atmosphere is filled with greenhouse gases. Progress, the novelist Ivan Klima once gloomily observed, is simply more movement and more rubbish. I think of the streets of Kathmandu.

My father may have been absent-minded and disorganized in some aspects of his life but he was remarkably shrewd when it came to property, even though as an academic lawyer he was never especially wealthy. When my family left Oxford for London in 1960 we moved to a huge Queen Anne terrace house, built in 1713, in the then run-down and unfashionable suburb of Clapham in south London. It was a very fine house with perfectly proportioned rooms, all wood-panelled and painted a faded and gentle green, with cast-iron basket fire grates (each one now worth a small fortune) in every room, and tall, shuttered sash windows looking out over the trees of Clapham Common. There was a beautiful oak staircase, with barley twist banisters. He had an eye for collecting antiques before it became a national pastime and impossibly expensive. So the new family home, with six bedrooms and almost forty windows – I painted them all once and then had a furious row with my father about how much he should pay me for the work – was filled with books, pictures and various objets d'art. I was immensely proud of all this when I was young. My father was also proud of his house and many possessions and liked to show them to visitors, but in an innocent and almost childlike way, wanting to share his pleasure with others. The family used to tease him that he was a wegotist, as opposed to an egotist – the word does exist in the *Oxford English Dictionary*.

My pride was of a more competitive and aggressive kind, albeit vicarious. When he eventually died at the age of ninety-six, my two sisters, brother and I were faced by a mountain of possessions. I discovered to my surprise that few, if any, of the many thousands of his books were worth keeping. It made me think about what would happen to all my books when I die. We divided everything else up on an amicable basis, but looking back I fear that I took more than my fair share, with my siblings acquiescing to their demanding younger brother so as to avoid disharmony. As for the house, with its forty windows and panelled rooms, I heard that it was recently sold for an astronomical sum, having been renovated. The estate agent's website showed the interior. It has been transformed: painted all in white, even the oak staircase, it now resembles an ostentatious five-star hotel.

When I am working in Nepal I live out of a suitcase, and have no belongings other than my clothes and my laptop. I have discovered that I do not miss my many possessions back in England at all – indeed I see them as something of a burden to which I must return, even though they mean so much to me. Besides, when I witness the poverty in Nepal, and the wretched effects of rapid, unplanned urbanization, I view my possessions in a different light. I regret that I did not recognize the virtues of trying to travel with hand luggage only at an earlier stage of my life. There are no pockets in the shroud.

'The first case is Mr Sunil Shrethra,' said the MO presenting the admission at the morning meeting. 'He was admitted to Norvik Hospital and then came here. Right-handed gentleman, sixty-six years old. Loss of consciousness five days go. On examination . . .'

'Hang on,' I cried out. 'What happened after he collapsed? Has he been unconscious since then? Did he have any neurological signs?'

'He was on ventilator, sir.'

'So what were his pupils doing?'

'Four millimetres and not reacting, sir. No motor response.'

'So he was brain-dead?'

The MO was unable to answer and looked nervously at me. Brain death is not recognized in Nepali law.

'Yes, sir,' said Bivec, the ever-enthusiastic registrar, helping the MO.

'So why was he transferred here from the first hospital if he was brain-dead?'

'No, sir. He came from home.'

I paused for a moment, unable to understand what this was all about.

'He went home from the hospital on a ventilator?' I asked, incredulous.

'No, sir. Family hand-bagged him, sir.' In other words,

the family took their brain-dead relative home, squeezing a respiratory bag all the time, connected to the endotracheal tube in his lungs to keep him oxygenated (after a fashion).

'And then they brought him here?'

'Yes.'

'Well, let's look at the scan.'

The scan appeared, shakily and a little dim, on the wall in front of us. It showed a huge and undoubtedly fatal haemorrhage.

'So what happened next?'

'We said there was no treatment so they took him home, hand-bagging him again.'

'Let's have the next case,' I said.

I had noticed that the sickest patients on the ITU, the ones expected to die or become brain-dead, had often disappeared by the next morning. I was reluctant to ask what had happened, and it was some time before I learnt that usually the families would take the patients home, hand-bagging them if necessary, so that they could die with some dignity within the family home, with their loved ones around them, rather than in the cruel impersonality of the hospital. It struck me as a very humane solution to the problem, although sadly unimaginable back home.

10

BROKEN WINDOWS

Back in Oxford, I went to inspect the lock-keeper's cottage. I walked with mixed feelings along the towpath, rain falling from a dull grey sky, past the line of silent narrowboats moored beside the still, green canal. The air smelt of fallen wet leaves. Several friends had told me that I was mad to try to renovate the place: after fifty years of neglect, with fifty years of rubbish piled up in the garden, without any road access, the work and expense involved would be enormous. The plumbing had all been ripped out by thieves for a few pounds' worth of copper, the plaster was falling off the walls, the window frames were all rotten. The ancient Bakelite electrical sockets and light switches were all broken. The roof was intact, but the staircase and many of the floorboards in the three small bedrooms were crumbling with woodworm. The old man who had lived there was dead, and the cottage itself was dead. The only life was the green wilderness of the garden, where the rampant weeds flourished after fifty years of freedom.

I had spent months making new windows in my workshop in London, with fanciful ogee arches. Glazing them with glass panes cut into ogee curves had been, therefore, difficult and time-consuming. With the help of a Ukrainian colleague and friend, I had ripped out the old windows and carefully installed the new ones before leaving for Nepal. While I was away in Kathmandu they had all been smashed by vandals. This was presumably out of spite for the metal bars I had fitted on the inside. As it was, the thieves had managed to prise apart the metal bars on the window at the back of the cottage and get in. At least I had put the more valuable power tools in two enormous steel chests with heavy locks that I had had the foresight to install. Wheeling them along the narrow towpath on a sack trolley had not been easy and at one point one of them, weighing almost 100 kilograms, had come close to toppling into the canal.

Apparently the thieves had mounted one of the chests on the sack trolley and then abandoned the effort as they couldn't open the front door – I had spent many hours fitting a heavy-duty deadlock to it. On the other hand, my elder sister, an eminent architectural historian, had remarked that the ogee arches were not very authentic for a lock-keeper's cottage; perhaps the vandals had shared my sister's rather stern views about architectural heritage.

I had therefore arranged for rolling metal shutters to be fitted on the outside walls over the windows, which completely defeated the original purpose of decorating the cottage with pretty arched windows. So the vandals then turned their attention, once I was away again, to the expensive roof windows – triple-glazed with laminated glass – that I had installed last year. They had climbed onto the roof, breaking many roof slates in the process, and then heaved a heavy land drain through one of the windows. As far as I could tell, this was done simply for the joy of destruction rather than for burglary – for the love of the sound of breaking glass. I consoled myself with the thought that the frontal lobes in the adolescent brain are not fully myelinated – myelin being the insulating material around nerve fibres. This is thought to be the explanation for why young men enjoy dangerous behaviour: their frontal lobes – the seat of human social behaviour and the calculation of future risks and benefits – have not yet matured, while the rising testosterone levels of puberty impel them to aggression (if only against handmade windows), in preparation for the fighting and competition that evolution has deemed necessary to find a mate.

Each time I walk towards the cottage I feel a sinking feeling at what further damage I will find. Will they have broken the little walnut tree or snapped off the branches of the apple trees? Will they have managed to break open the metal shutters? In the past I always felt anxious when my mobile phone went off for fear that one of my patients had come to harm. Now I fear that it will be one of my friendly neighbours from the longboats nearby on the canal or the police, informing me of another assault on the cottage. I tell myself that it is absurd to worry about mere property, especially as the cottage only contains building tools, all locked up in steel site chests. I remind myself of what I have learnt from my work as a doctor, and from working in poor countries like Nepal and Sudan, but despite this the project of renovating the cottage has started to feel like a millstone. It fills me with a sense of despair and helplessness, when I had hoped it would give me a sense of purpose.

In the weeks before I left for Nepal I had started to clear the mountains of rubbish from the garden. At one end of the garden there is a brick wall, and on the side facing the canal there was a mass of weeds and brambles. I had cleared these to reveal a series of picturesque arched horse troughs made of red brick. The bricks had been handmade – you could see the saw marks on them. They would have been for the horses that pulled the barges along the towpath in the distant past, and there were rusty iron rings set into the bricks for tethering the horses. In front of the troughs, and still on my property, was a fine cobbled floor, which slowly appeared as I scratched away years of muck and weeds. Emma, one of the friendly boat people, stopped by to chat as I worked.

'There is a rare plant here,' she said. 'The local foragers were very excited, though I'm not sure what it's called. Fred and John [two other local boat dwellers] got into trouble with them a few years back when they tried to clear the area.'

'I'm worried that I might have dug it up,' I replied, anxious not to fall out with the local foraging community.

'Oh it will probably grow back,' she said. 'It has deep roots.'

We talked about the old man. He had been frightened of thieves, Emma told me, although as far as I could tell from the rubbish, he had owned little and lived off tinned sardines, cheap lager and cigarettes. He had also told her that the cottage was haunted. According to the locals he had been 'a bit of a wild one' when he was younger, but all I got to hear were stories of how he would sometimes come back to the cottage drunk on his bicycle and fall into the canal. He had a son who had once lived in the cottage with him for a while, but it seems that they had become estranged. There had been a few pathetic and broken children's toys in the rubbish in the garden. I had found shiny foil blister packs of antidepressants – selective serotonin reuptake inhibitors – in the piles of rubbish in the garden. Emma told me that he had died in the cottage itself.

'We didn't see him for several days and eventually got the police to break the door down. He was very dead – in an armchair.'

I slowly built up a huge mound of several hundred black plastic builders' bags, filled with fifty years of my predecessor's rubbish and discarded possessions. It included a matted pile of copies of the *Daily Mail* that was almost three foot thick and had acquired the consistency of wood, having been exposed to the elements for a long time. Rusted motorcycle parts, mouldy old carpets, plastic bags, tin cans, bottles galore (some still containing dubious-looking fluid), useless and broken tools, the pathetic children's toys – the list was endless. None of the rubbish was remotely interesting; even an archaeologist excavating it five hundred years from now, I thought as I laboured away, would find it dull and depressing. The more I dug down, the more rubbish I found.

The community of boat dwellers along the Oxford Canal is supplied with coal and gas cylinders by a cargo barge called *Dusty*. When I took possession of the cottage I found a cheerful note put through the letterbox from Jock and Kati, the couple who own *Dusty*, welcoming me to the cottage and offering me their services. This proved invaluable because with their help I was able to load two bargeloads of rubbish onto *Dusty* and take it up the canal a short distance to where a local farmer had agreed I could put a couple of big skips beside a farm track. It was heavy work, and when it was done I took Jock and Kati out to lunch in a nearby pub. Jock told me that he had backpacked round the world and then become an HGV driver, but he had always wanted to live in a boat from an early age. Kati was a primary school teacher who had taken a year off work and was now reluctant to return. They spent the day travelling slowly up and down the canal, delivering sacks of coal and cylinders of gas to the boat dwellers, all of whom they knew. It was like living in a village. They were very happy, they told me, with their slow and peaceful life, uncluttered by possessions, living in a second barge moored further along the canal.

I had to fell several trees – mainly thorn trees, over thirty feet tall – which had taken over one corner of the garden. Much as I love trees, to the point of worship, I must confess that I also love felling them and I own several splendid chainsaws. After some years I have finally mastered the art of sharpening the chains myself. I suppose tree surgery has a certain amount in common with brain surgery – in particular, the risk and precision. If you don't make the two cuts on either side of the trunk in precisely the right place the tree might fall on you, or fall to become jammed in the surrounding trees, which makes further work extremely difficult; or the bar of the chainsaw can get completely stuck in the tree trunk. And the chainsaw must be handled with some care – I once saw a patient whose chainsaw had kicked back into his face. But there is also the smell of the cut wood – oak is especially fine – mixed with the chainsaw's petrol fumes and, depending on where you are working, the silence and mystery of being in a forest. One of the first books I read as a child – perhaps because my mother was German – was *Grimms' Fairy Tales*, with its many stories of devils, bloody death and punishment, set in dark woods. Felling trees is also a little cruel – like surgery. There is your joy in mastery over a living creature. To see a tall tree fall to its death, especially if you have felled it yourself, is a profoundly moving sight. But what makes brain surgery so exciting is your intense anxiety that the patient should wake up well, and you fell trees to provide wood for making things or for firewood, or to help the growth of other trees. And, of course, you should always plant new ones.

Twenty-five years ago I acquired twenty acres of land around the farmhouse in Devon where my parents-in-law from my first marriage lived. I planted a wood of 4,000 trees in eight acres – native species, oak and ash, Scots pine, willow and holly. For a few short years I could happily tend the trees when I visited Devon, carefully pruning the lower branches of the young oaks, so that after a hundred years they would provide long lengths of knot-free, good-quality timber. I made an owl box and put it up in the branches of an old oak tree growing in one of the hedges that lined my land. I once saw an owl sitting thoughtfully in the box's large opening, which was a very happy moment, but to my disappointment the owl did not take up residence in it. I hoped that I would be buried in the wood after my death, and that eventually the molecules and elements of which I am made would be rearranged as leaves and wood. I had no idea at all of the disaster that awaited my marriage. I lost the land and the trees with divorce, and they were soon sold off. You can still see the wood, now overgrown and neglected, on Google Earth. A third of the trees should have been felled to allow the remaining ones to grow stronger, but this has not been done.

I miss the place greatly – not only the fields and the wood, but the workshop I set up in one of the ancient cob-built barns opposite the farmhouse. The windows, which I had made myself, in front of the workbench, which I had also made, looked out over the low hills of north Devon towards Exmoor. Swallows nested in the rafters above my head, and the young ones would learn to fly by fluttering from beam to beam. Their parents would dart in through the open doors and, if they saw me, would at first turn a somersault directly in front of me – I could feel the air under their wings in my face – and then shoot out again, but after a while they became used to my presence. By late summer the young birds would be flying outside in the farmyard and gather on the cables that stretched from the farmhouse to the barns, little crotchets and minims, making a sheet of sky music. Before autumn came, they would leave for Africa. I returned to have a look at the farm twenty years later, explaining to the new owner my connection with the place. He proudly showed me all the improvements that he had made. I probably should not have gone: the barns and my workshop had been converted into hideous holiday chalets and the swallows evicted, never to return.

Once I had cleared the tons of rubbish from the garden of the lock-keeper's cottage, I planted five apple trees and one walnut tree. The apple trees were traditional varieties such as Cox's Orange Pippin and Blenheims – the same kind of trees as those in the orchard of my childhood home nearby, where I had grown up sixty years earlier.

It had been a working farm until only a few years before my father bought it in 1953, when I was three years old. It was a very fine Elizabethan stone building with a stone roof. There was a farmyard with thatched stables, a large pantiled barn and the orchard and garden, with sixty apple and other fruit trees and a small copse – a paradise, and an entire world for a child such as myself. It was on the outskirts of Oxford, where open fields met the city. There is now a bypass running over the fields. The neighbouring farm has been replaced by a petrol station and hotel. Most of the orchard has been felled, and a dull housing estate has replaced it. The barn and stables have been demolished. There is still a pine tree there, stranded among the maisonettes and parked cars, which had stood guard at the entrance to the copse. I was frightened of the copse, and thought it was full of the witches and devils I loved reading about in fairy stories. I remember how I would stand by the pine tree, sixty years ago – the tree must have been much smaller then, but looked enormous to me. I was too scared to enter the deep and dark forest it guarded, despite longing to be a brave knight errant. Sometimes, as I stood there, I could hear the sound of the wind in its dark branches above me, and it filled me with a sense of deep and abiding mystery, of many things felt, but unseen.

We had many pets, one a highly intelligent Labrador called Brandy. He belonged to my brother but I wanted to train him to sit and beg. I'm not sure why – I now hate to see animals trained to do tricks. But I did it with great cruelty, using a whip made from electric cable, combined with biscuits. He learnt quickly, and I enjoyed the feeling of power over him until my mother found me once with the poor creature. The dog would never stay alone with me in a room for the rest of his life, a constant reminder of what I had done, however much I now tried to persuade him of my love for him. I was filled with a deep feeling of shame that has never left me, and a painful understanding of how easy it is to be cruel. This was also an early lesson in the corrupting effect of power and I wonder, sometimes, if this has perhaps made me a kinder surgeon than might otherwise have been the case.

I had a slightly similar experience when I started work as an operating-theatre porter in the northern mining town. There was an elderly anaesthetist who I now realize was appallingly incompetent. On the first day that I was on duty to assist him, he seemed to be having difficulties intubating the patient, who started to turn a deep-blue colour (known as cyanosis, the consequence of oxygen starvation). In all innocence, I asked him if patients normally went blue when he anaesthetized them. I do not remember his response – but the other theatre porters fell about laughing when they heard the story. A few weeks later he was having difficulties intubating another patient, who started to struggle – the poor man clearly had not been anaesthetized properly. He told me to hold the patient down, which I did with enthusiasm. I had always liked a good fight when I was at school (although there are more shameful episodes there as well, when my strength and aggression got the better of me and my schoolmates started crying). At that point Sister Donnelly, the theatre matron, entered the anaesthetic room and saw how I was restraining the patient. 'Henry!' was all she said, looking genuinely shocked. I cannot forget it. Perhaps it was these experiences that make me cringe when I sometimes see how other doctors can handle patients.

When I worked as a psychogeriatric nursing assistant many years later, it was obvious that the atmosphere on each of the wards was largely determined by the example set by the senior nurses in charge, most of whom understood the duty of care, and how difficult it can sometimes be, as it was a real and daily obligation. As authority in hospitals has gradually passed from the clinical staff to non-clinical managers, whose main duty is to meet their political masters' need for targets and low taxes, and who have no contact with patients whatsoever, we should not be surprised if care suffers.

At my home in Oxford, with its ancient house and garden – my little paradise – I ran a bit wild, the spoilt youngest child of a family of four. When we moved to London when I was ten years old, it was as though I had been evicted from the Garden of Eden.

As I walked along the towpath towards the cottage I also thought about why I had bought it in the first place and why I felt the need to do it up myself. Most of my life was behind me and I found the physical work involved increasingly difficult and much of it positively depressing. Work seemed to be going backwards, not forwards, let alone the damage caused by the vandals. When I cut into the plaster to install a new power socket, huge pieces of it fell off the wall. The lath-and-plaster ceiling of the room downstairs collapsed in a cloud of dust when I tried to strip the polystyrene tiles that had been glued to it. The new windows that I had made myself had all been smashed and I would have to reglaze all of them, and now one of the roof windows as well. Besides, if I ever finished the work, what would I do then? I had to conclude that what I was doing was not just to prove that I was capable of such work despite growing old, but also an attempt to ward off the future. A form of magic, whereby if I suffered now, I would somehow escape future suffering. It was as though the work involved was a form of penance, a secular version of the self-mortification found in many religions, like the Tibetans who crawl on all fours around Mount Kailash in the Himalayas. But I felt embarrassed by the way in which I was doing all this in the cause of home improvements – it seemed a little fatuous when there was so much trouble and suffering in the world. Perhaps I am just a masochist who likes drawing attention to himself. I always was a tremendous show-off.

Thinking these depressive thoughts, I arrived at the cottage but, just as on my previous visits, as soon as I saw it I had no doubts. There was the wild garden and the old brick horse troughs with the quiet canal in front and the lake behind, lined on one side with tall willows. The two swans were there, perfectly white on the dark water and, beyond them, reeds faded brown with the winter, and then the railway line, along which I had once watched steam trains roaring past as a child. When I went inside – in darkness, now that the broken windows were all boarded up – the light from the open door fell on broken glass, shining and scattered all over the floor, which crunched underfoot as I entered. But it no longer troubled me.

I would restore this pretty and humble building, I would exorcize the old man's death and all the sad rubbish he had left behind. The six apple trees and one walnut tree would flourish. I would put up nesting boxes in the trees, and an owl box, like the one I had installed in the old oak tree in the hedge beside the wood in Devon.

I would leave the cottage behind, for somebody else to enjoy.

I decided to put motion-detecting floodlights high up on the cottage walls, and also CCTV cameras – a reluctant concession to the thieves and vandals. This involved working up a ladder, high under the eaves of the roof. I have lost count of the number of elderly men I have seen at work with broken necks or severe head injuries sustained by falling off ladders: a fall of only a few feet can be fatal. And there is a clear connection between head injuries and the later onset of dementia. I therefore drilled a series of ringbolts into the cottage wall, like a climber hammering pitons into a rock face, and tied the ladder to the ringbolts and, wearing a safety arrest harness, attached myself to the ladder with carabiners as I fixed the lights and wretched CCTV cameras, wielding a heavy-duty drill to bore through the cottage walls for the cables.

While I was doing this work I received a visitor. I climbed down the ladder. He was a man my age, walking with a golden retriever, which happily explored the wild garden while we talked.

'I lived here as a child sixty years ago,' he said, 'in the 1950s, before Dennis the canal labourer took it over. My brother and I lived here with our parents. It was the happiest time of their lives.'

We worked out that we were of the same age and had lived at the same time in our respective homes less than one mile apart. He produced an old black and white photograph showing the cottage looking tidy and well cared for, with a large flowering plum tree in the front garden. You could just see that the garden had many vegetables growing in neat rows. His mother was standing at the garden gate, wearing an apron.

'I scattered my parents' ashes over there,' he told me, pointing to the grassy canal bank on the other side of the little bridge across from the cottage. 'I come here to talk to them every so often. I told them today that their grandson had just got a university degree. They would have been so proud.'

I showed him around the inside of the cottage. He gazed at it in silent amazement – so many memories must have come back.

'My dad used to sit in the corner over there in the kitchen,' he said, pointing to the place where there had once been a stove. 'He had a handful of lead balls. He'd throw them at the rats when they came in through the front door, but I don't know if he ever got one.'

11

MEMORY

By the time that he died at the age of ninety-six my father had become profoundly demented. He was an empty shell, although his gentle and optimistic good nature remained intact. The live-in carers my brother had organized to look after him in his flat often remarked on how easy it was to look after him. Many of us, as we dement, increasingly confused and fearful as our memory fades, become aggressive and suspicious. I had seen this myself when I worked briefly as a geriatric nursing assistant, although in the grim and hopeless environment of a long-term psychiatric hospital – an environment which must have made the poor old men's problems many times worse than they might otherwise have been. He was famously eccentric – the porters at the Oxford college where he had been a don after the Second World War regaled me – when I went there as an undergraduate myself many years after he had left – with stories of his many eccentricities. He once met one of his former pupils, who told him that he had always been terrified when having a tutorial with him. My father, the mildest of men, was painfully surprised, until his former pupil went on to explain that he had rarely had any matches with him when giving tutorials. He would light the gas fire in his room by turning on an electric fire – one of those old models with a red-hot bar – and pressing it against the gas fire. The tutorial would therefore start with an alarming explosion. There had been an electric fire like that in my bedroom in the old farmhouse. There was no central heating: in winter there would often be frost flowers on the bedroom windows in the mornings and I would lean out of bed, trying to stay under the blankets, heating my clothes in front of the fire before getting up to bicycle to school.

I used to read late at night with a torch under the blankets after my mother had kissed me goodnight and turned off the lights in the room. At the age of seven I borrowed a school friend's book about King Arthur and his knights. I became slightly obsessed by these stories and read everything I could find about knights and chivalry, including Malory's *Morte d'Arthur*. I considered Lancelot and Galahad to be hopeless goody-goodies but greatly admired Sir Bors, who was tough, loyal and reliable. He would have had no time for women or religion, I thought. My edition of Malory had many coloured illustrations by Sir William Russell Flint, the popular late-nineteenth-century artist, who was famous for his erotic paintings of women. His illustrations for Malory had heroic knights and beautiful maidens with long Pre-Raphaelite hair in tresses and wearing long and flowing robes, which I found highly attractive. This night-time reading probably contributed to my severe short-sightedness, which resulted in retinal detachments many years later.

I found my weekly attendances at Sunday School extremely boring, and the illustrations in the little books of Bible stories for children very dull compared to the pictures in the *Morte d'Arthur*. Both my parents were sincere – although relaxed – Christians. I received a traditional English middle-class Christian education at Westminster School, including morning service six days a week in Westminster Abbey when I was a teenager. Every so often the organist would play the last movement of Widor's Fifth Symphony – the only piece of French organ music I could abide – at the end of the service. I would stay behind in the now empty building as the music crashed and boomed under the great Gothic roof and round all the marble statues and monuments, until my anxiety about being late for the first class would overcome me and I would run back to the school through the empty cloisters, over the worn gravestones, with the music fading behind me.

I was bitterly unhappy in my first year at the school. I was a boarder for the first time in my life. I think my parents thought it would be good for me, and it was a fairly traditional part of a middle-class boy's education at the time. I missed having my own room and, being very innocent and prudish, was shocked by the other boys' endless talk about sex. I once went to the housemaster to complain about this – I squirm at the memory. After a year I finally dared to tell my parents how unhappy I was. I remember the overwhelming sense of relief when I realized that it was going to be possible for me to become a day boy.

In my last year at the school I spent Friday afternoons in the Abbey Muniments Room, filing nineteenth-century inquest reports from the Westminster Coroner's Court. The Cadet Corps had been abolished. Until then we had spent Friday afternoons in military uniforms marching round the school yard with ancient rifles, .303 Lee Enfields said to date back to the Boer War, but converted to .22 bore. We were offered various alternatives to the Corps and I chose the Muniments Room. The room was part of the south transept of the Abbey, above the aisle, and you looked directly down into the Abbey itself. I was tasked with filing a large number of

Coroners' Inquests from the Westminster Court in the 1860s. The reports, bound in crumbling green tape, were kept in a huge, semicircular medieval chasuble chest made of oak which was black with age. I found an old sword on top of the chest. Henry V's sword, I was told, which indeed it was. I liked to wave it around my head while quoting the appropriate lines from Shakespeare. The Keeper of the Muniments was a small, round and bird-like man who wore bright yellow socks and rolled rather than walked. He took very long – probably liquid – lunch breaks so I was left largely to my own devices. Although I found the inquests fascinating – stories of death in Dickensian London in perfect copperplate writing – I was more excited to discover a spiral stone staircase leading up from the Muniments Room to the triforium and roof of the Abbey. I therefore spent most of my Friday afternoons exploring all the empty spaces and the roof of Westminster Abbey, with wonderful views of central London.

As far as I can remember, I never believed in God, not even for a moment. At one morning service in the Abbey I remember seeing the school bursar – a retired air commodore – praying. He was kneeling opposite me on the other side of the gilded choir stalls. There was a look of the most terrible pain and pleading on his face. He disappeared from the school shortly afterwards and I heard later that he had died from cancer.

My father's dementia was probably avoidable. He had suffered two significant head injuries in his seventies – once when falling between the rafters of a friend's attic and knocking himself unconscious on a marble fireplace in the room below, and once falling off a ladder when trying to read the gas meter in his huge eighteenth-century house in London. He already had form for losing his footing between attic rafters: in the ancient house in Oxford where we lived in the 1950s, his leg, much to the surprise of the family's au pair, once appeared in a shower of plaster through the ceiling above her bed, but fortunately not the rest of him. At the time of the two head injuries he had seemed to make a reasonable recovery, but they probably contributed to his slow deterioration in old age.

I was not a good son. In his declining years, after our mother's death, although I lived quite nearby I rarely went to visit him. I was impatient with his forgetfulness and distressed by the fact that he was no longer the man that he had been. My siblings went to see him more often than I did. Both my parents expected very little from me – they took pleasure in any success that I had, and were always anxious to help me in any way that they could – yet rarely, if ever, seemed to ask for anything in return for themselves and rarely, if ever, complained. I exploited their love, although their love was certainly the principal source of my feeling of self-importance, something which has been both a strength and a weakness throughout my life.

My father had been an eminent lawyer, although his career is not easy to categorize. An Oxford don for fourteen years, he left Oxford to run various international legal organizations and finally worked for the British government on reforming and modernizing British law as one of the first Law Commissioners. When I was young I had no interest in the law or in my father's work – it seemed terribly boring. It was only in the closing years of my own career as a doctor, mainly from my work overseas, and having to witness so much corruption and the abuse of power, that I came to understand the fundamental importance of the rule of law to a free society – the principle that lay at the heart of my father's work and view of the world. Democratic elections, for instance, mean little without an independent judiciary. His obituary in the London *Times* filled an entire page and I was filled in turn with both filial pride and guilt. My own career, as a doctor, now seems to me rather slight in comparison to his.

The profoundly serious nature of his work – and his deeply moral and almost austere view of the world – were completely at odds with the way he did not seem to take himself at all seriously. The family followed his lead, and I fear that we treated him more as a figure of fun than of authority. There were only a few rare occasions when he would briefly lose his temper with us over the way that we treated him with such a singular lack of respect. He enjoyed telling stories against himself and of his – not entirely unselfconscious – eccentricities. He often talked of writing his memoirs but he never got beyond the first page, which described how he pulled the lanyard on an artillery piece in Victoria Park in Bath in 1917, at the age of four, as a reward for his parents buying war bonds to help finance the First World War. I often said that I would sit down with a tape recorder and record his many memories and stories – he had led a most unusual and very interesting life and was an excellent raconteur – but I never did, and this is something I deeply regret. His past, and many of the stories that he had heard in turn about his family's origins in the countryside of Somerset and Dorset, faded as his brain decayed and are now lost for ever. I know only a few fragments.

During the war he had been in Military Intelligence, interrogating high-ranking German prisoners of war, as he spoke German. His preferred technique, he once told us, was to treat the interrogation sessions like an Oxford tutorial, and encourage the prisoners to write essays for him on democracy and the rule of law. 'The hardened Nazis were a lost cause,' he said. 'But it worked for some of the others.' One U-boat captain, he discovered, was something of an anti-Nazi, so my father dressed him in a British Army great-coat and smuggled him out of the prison camp to take him on a tourist trip round London, although he said he was a little worried as to what he would say if they were stopped by the police. He was outraged when stories emerged of the British Army hooding – effectively torturing – IRA suspects in Northern Ireland at the start of the Troubles. Like many experienced interrogators, he was of the opinion that kindness and persuasion worked much better than torture. His particular interest, when interrogating prisoners, was in German morale. He wrote a report arguing that the carpet bombing of German cities was strengthening it rather than breaking it. Apparently, when 'Bomber' Harris – the head of RAF Bomber Command – saw the report, he was so enraged that he wanted to have my father court-martialled. This did not happen, fortunately, and history, of course, has entirely vindicated my father.

He liked to claim – I suspect with some exaggeration – that there were only three books in his parents' house in Bath, where his father ran a jewellery business and his mother had a dress-making business until she had children. She was a farmer's daughter – one of eight children – and used to walk eight miles to work every day as a seamstress, eventually owning her own shop, which, our father assured us, had been highly fashionable. He had a difficult relationship with his father, and once had even come close to blows.

'Did you feel bad about that?' I asked him, when he once told me this.

'No,' he replied calmly. 'Because I knew I was right.'

His certainty was not arrogance, but was based on a coherent and stubbornly moral outlook which was usually correct, and which was most frustrating for the rebellious, selfish teenager that I once was. His own father – whom I never knew, as he died shortly after my birth – could not understand how his son had become a left-leaning liberal intellectual, with a house packed with many thousands of books and a German refugee for a wife.

Before reading law at Oxford he had been educated at a minor public school near Bath which, he once told me, specialized in turning out doctors and evangelical missionaries. In later life he said he still had nightmares about the place – *Tom Brown's Schooldays* were nothing, he once told me, compared to what he had to put up with. So he hated the place, yet he became the Sergeant Major for the school's army cadets, played rugger in the First XV and rowed in the First Eight. He said that he owed everything that he became to one inspirational history teacher. His attitude to success and conventional authority remained deeply ambiguous throughout his life. One of the few times I saw him seriously distressed (other than the various occasions when I caused him great pain) was when the teacher who had so inspired him wrote to him asking for a contribution to a fundraising campaign for his old school. My father was a deeply generous man and involved in much charitable activity, but after a great deal of painful heart-searching, he wrote to his old teacher and mentor saying that he felt unable to send any money.

He had been secretary to the Oxford League of Nations Association – an organization both idealistic and doomed – and had met my mother when he went to Germany to learn German in 1936. He stayed in the town of Halle, where he found himself in the same lodgings as my mother, who was training to be a bookseller. Her strongly anti-Nazi political views had prevented her from going to university and so she chose bookselling as the career closest to her love of philology. My father was the first person she met to whom she could pour out her heart about her deep unhappiness as to what was happening in Germany. Her views eventually got her into trouble with the Gestapo. Her colleagues at work, with whom she had been overheard exchanging anti-Nazi views, were tried and imprisoned but my mother was let off on the grounds that she was – as one of the two Gestapo men who interrogated her put it – a 'stupid girl'. They told her, however, that she would be cross-examined as a witness in her colleagues' trial and she felt that she would not be able to cope with this. So, in brief, my father married her and brought her to England, a few weeks before the Second World War started.

My mother's sister was an enthusiastic supporter of Hitler and the Nazis, and her brother joined the Luftwaffe, although from a love of flying and not from any political conviction. I do not know how my mother knew that Hitler's regime was evil. It is only since her death, by reading about Germany at that time (in translation, given my shameful lack of German) that I have come to understand just how remarkable was her defection. Her decision to leave Germany – a country with the deepest respect for authority and on the brink of war – would have been seen by many as treason. It seems obvious and easy in retrospect, but how I wish she was still here so that I could talk to her about this.

Sixty years later, I asked my parents about their decision to get married. My first marriage was falling violently apart, and sometimes I went to speak to them about my unhappiness. I should not have burdened them with this, but it was a strange experience to converse with them both almost as equals, as fellow adults, about the difficulties of married life. I learnt that the decision to marry my mother had been a difficult one for my father, although he did not specify exactly why. My mother suggested he was on the verge of marrying somebody else in England, but my father did not confirm this. He told me that he was in such a state of despair that one day, working as a young lawyer in London, he was walking up Tottenham Court Road and saw a doorway with a sign advertising a counselling service. I think it was some kind of Christian mission – whatever it was, my father said that he went in and found a man there who helped him greatly.

In Nepal marriages are still arranged. My Nepalese friends tell me that it usually works. Sometimes it seems to me that my parents' highly successful marriage was, in a way, also arranged – arranged on the basis of the rule of law, morality and liberal democracy. They were closely involved from the 1960s onwards in the creation of Amnesty International, and my mother ran – in her quiet and wonderfully efficient way – the registry of political prisoners in dozens of countries. At the beginning the offices were in the chambers of the lawyer Peter Benenson, who had come up with the idea for the organization. I would go in sometimes to help – mainly to lick and stamp envelopes. I would like to say these were for letters sent to dictators all over the world, but they were mainly newsletters to the small groups of volunteers – organized in cells like revolutionaries – who adopted particular prisoners and would write in protest to the dictatorial regimes that had imprisoned them.

By the time that I was born in 1950 my mother had developed a strange, disabling condition which resulted in excruciatingly painful bruising over many of her joints. She consulted a wide number of specialists but none could come up with a diagnosis. One suggested it was allergic, so they had the family pets put down. The only treatment that seemed to help was arsenic (from which she developed a rare skin cancer called Bowen's Disease many years later).

Eventually in despair – I think I was only a few months

old at the time – they sought a psychiatric opinion and my mother was admitted for six weeks' inpatient psychoanalytic treatment in the Park Hospital in Oxford under the care of a fellow German émigré. He was, she once told me, very much like her father, who had died quite suddenly in 1936 from metastatic bowel cancer when she was nineteen. As a fellow émigré, he must also have understood her deep sense of loss – of her family, of her past and much of her identity. Her mother had died from breast cancer during the war and her sister perished in a British air raid on the city of Jena. Her sister had been an enthusiastic Nazi and she and my mother had parted on bitter terms when my mother left for England, although my mother was told after the war that her sister had changed her mind before her death. And perhaps, although I never asked her directly, she also felt that she had betrayed her principles by fleeing from Germany and by not standing up for them and her colleagues in court.

The treatment worked and the stigmata-like bruises disappeared. There is no way of knowing whether it was the psychoanalysis which helped, or whether it was the way in which my father, by his own admission, became a more considerate husband because of her illness, or whether it was simply the rest in hospital from bringing up four children with minimal help and from a husband who was completely dedicated to his work. My parents once told me – as a joke – that my own personality, which at times caused them great problems, was partly to be explained by the famous child psychologist Bowlby's work on maternal deprivation. This may or may not be true, but only as I reach old age myself have I come to understand just how completely I am my parents' creation, and whatever good is in me came from them.

Thirty years later, by which time I was a medical student, my mother's purpuric swellings, or 'bumps' as she called them, reappeared, much to my parents' alarm. 'You must be stressed and anxious about something!' I remember my father saying to her, close to despair, as she lay on her bed in severe pain. But on this occasion a specialist prescribed the drug dapsone – a drug normally used for leprosy – and the bruises immediately disappeared. I still do not know what the underlying diagnosis might have been. If dapsone, a mere chemical, had been available in 1950, perhaps I would not be the person I am now, and I might not be sitting in a remote Nepalese valley, at the foot of Mount Manaslu, writing about my parents as I listen to the water of the glacial River Budhi Gandaki, rushing noisily past over its many rapids on its journey from the Himalayas to join the River Trishuli and then the Ganges, to end in the Indian Ocean.

My father was a tremendous optimist – even as his memory deteriorated he would express the hope that things would get better. Before his dementia became profound, while he was still living in the grand eighteenth-century house overlooking Clapham Common, I once engraved a brass plate with the family name to go by the bell push for the front door to distinguish it from the bell for the basement flat, which was now rented out. My engraving was very clumsy and the letters became smaller as they went from left to right. 'Like my faculties,' he said sadly as we looked at it together after I had screwed it in place. He still had some insight then, into what was happening to him. 'But I think things will get better.'

On my second trip to Nepal I accompanied Dev to a Health Camp he had organized in a remote corner of the country. The metalled road ended at Gorkha, and we then took three hours to travel the thirty-six kilometres over the mountains to the small town of Arughat on a wildly uneven dirt track – though not uneven enough to dissuade large trucks and buses from crawling along it as well, throwing up huge clouds of ochre dust. In places there was barely room for the vehicles to squeeze past each other, often with a precipitous drop, only a few centimetres away, to the valley below. On a clear day we would have seen Mount Manaslu at the head of the valley, one of the most beautiful of the great Himalayan peaks and the eighth-highest mountain in the world, but the haze was intense and there was no view at all.

The Health Camp took place in what had been a brand-new primary care hospital, but which had been badly damaged in the earthquake a few days before it was due to open. It had been abandoned until Dev had come to inspect it and found that most of the building was serviceable, although in a terrible mess. An advance party had arrived a day before us and I was very surprised to find a clean and tidy building – although with great cracks in the walls and one wing partially collapsed – when we arrived. The Neuro Hospital team of over thirty doctors, nurses and technicians had come with enough equipment to run two operating theatres, a pharmacy and five outpatient clinics, with plain X-rays and ultrasound and a laboratory. It was an impressive piece of organization – but they had done similar camps elsewhere before, especially after the earthquake, and had learnt from experience.

Next morning there was a long queue of patients – several hundred, mainly women, all dressed in brilliant red – waiting outside the hospital. All the treatment would be free. They were sheltering under equally colourful umbrellas, as the temperature was soon in the 90s. They were held back at the hospital gates by armed policemen and allowed in, one by one, to be registered at the entrance and directed to the appropriate clinic – such as orthopaedic, plastic-surgical or gynaecological. Fifteen hundred patients were seen in three days and many, relatively minor, operations performed, some under general anaesthetic. Difficult cases were advised to go to larger hospitals long distances away. Patients came from far and wide – the Health Camp had been advertised for many days in advance.

'Some patients are coming from the Tibetan border,' I was told.

'How far away is that?'

'Four or five days' walk. No roads. Ten days if you or I tried to do it.'

Sick patients arrived on stretchers. Several old people arrived carried piggyback.

Although I was treated like visiting royalty, and presented with bougainvillea garlands and silk scarves at the lengthy opening and closing ceremonies by the local people, and a framed certificate entitled 'Token of Love', I was completely useless. My days of general surgery and general medicine are long behind me. I was disappointed to find, as I watched Dev happily operating on inguinal hernias, hydrocoeles and similar lesions, that I had completely forgotten how to do them, even though I had spent a year doing dozens of such cases when working as a general surgeon. Thirty-five years ago, a year of general surgery had been a necessary part of qualifying for the final FRCS examination, which I had to pass before I could train as a neurosurgeon. I sat in on some of the Health Camp clinics, and I found that the junior doctors knew much more than I did.

Dev is doing a clinic – a chair has been put next to him for me. A tidal wave of patients now flows in: an old woman with elaborate gold ornaments in her nose and a rectal prolapse, old men with inguinal hernias, old women with haemorrhoids, many patients with varicose veins – cases which remind me why I was pleased when my year of general surgery came to an end almost forty years ago, and I could devote myself to neurosurgery. But it also reminded me how modern medicine is not just about prolonging life – it has probably achieved as much good by finding treatments for all the chronic non-fatal conditions from which we would otherwise suffer and from which people in poor countries like Nepal still suffer.

I remembered a rectal clinic I had done on Friday afternoons when I did my mandatory year of general surgery. Neither I, nor my patients, enjoyed the experience. They knew, and I knew, that I was going 'to give them a ride on the silver rocket' – the medical procedure of sigmoidoscopy, where an illuminated long stainless steel tube is used to examine the inside of the rectum. But I was happy enough when I was in the operating theatre.

There is a young woman with unilateral proptosis – her right eye is bulging outwards; we will send her to Kathmandu for a scan. There is a girl with what are probably pseudo-seizures, hurried in by her anxious mother. If people have fits in front of the doctor – which is what is happening here – it usually means, although not always, that the problem is psychological rather than epileptic. Dev prescribes the antidepressant amitriptyline. There's no question of any follow-up in such a remote country area. It is impossible to know what will happen to the patients. Many of them – most of whom are illiterate – produce plastic bags full of the many medicines they have been taking.

It's all in Nepali, of course. I am half asleep, lulled by the sound of hundreds of voices outside and the whirring of the ceiling fan. The temperature outside must be in the high 90s. The patients at the front of the queue are pressed up against the metal gates, and the police guards push themselves into the crowd from time to time to stop fights breaking out, or to allow urgent cases to be brought into the hospital. But inside the building everything is highly organized.

There is a man with huge, wart-like growths on his hands and feet. Next we see a five-year-old boy and his ten-year-old sister, who both went blind at the age of two. They are led into the room and sit sightlessly while my colleagues thumb through the stained and dog-eared pieces of paper that comprise their medical notes. All we can do is confirm that there is nothing to be done. I ask whether there are schools for blind children in Nepal and am told that there are, but that it is unlikely that these children, from a remote mountain village, could go to one.

At lunch on the baking-hot roof, sitting in the shade of a bright-blue UNHCR tarpaulin left over from earthquake relief, I talk to the gynaecologist.

'How many PVs [vaginal examinations] have you done so far?'

'Over five hundred.'

'Do the women know any anatomy?'

'Most know none at all. It's a waste of time trying to explain anything to them. A few of them can understand. But usually I just say take the medicine and this is the name of your illness. The women in the queue outside the room,' she adds, 'were starting to fight each other, trying to get in . . .'

One room is reserved for people who are too weak to sit or stand. There is a young woman with diabetes who is now in severe ketoacidosis. She lies on a stretcher, with dulled eyes and a deeply resigned expression on her face, coughing and retching into a plastic bowl from time to time. The MO fails to get a drip up on her and I reconfirm my uselessness by also failing. One of the anaesthetists succeeds. We give her IV fluids and find some insulin from another patient.

'What's her outlook?' I ask.

'Not good. Poor peasant in a remote village. Can't afford insulin. Diabetes is still a fatal disease here for many people. But we've told her husband to take her to the nearest big hospital. They may be able to help.'

I find an empty room in the ruined part of the hospital. There are large cracks in the wall from the earthquake. The many windows look out onto some tall mango trees and the room is full of the sound of the wide River Budhi Gandaki rushing past, coming down from the glaciers of the invisible Mount Manaslu. I sit there quietly for a while, trying to write, until two playful Nepali boys find me and peer over my shoulder at what I am doing and will not leave me alone, so I return to the clinic to watch more patients come and go.

After three days, Health Camp comes to an end. By the early evening there are only a few patients left waiting outside the entrance. I sit outside on a white plastic chair looking at the dim, blue hills around me. It is still very hot, and there is a strong wind, so the giant mango trees wave and shake. A noisy wedding party passes on the nearby road, dust swirling about them, the women all brilliantly dressed, with two men at the head of the procession blowing on long, curved horns. The bride is carried in a palanquin, and is veiled and dressed in red and flashing gold. The groom walks behind, his face heavily made up, wearing an elaborate and decorated coat. Three young girls are playing in the hospital courtyard and come up to me. We exchange a few words, which we don't understand, and they happily laugh and dance around me for a while, before running away. I so wish I spoke their language and could talk to them. My inability to speak any language other than English is the deepest of all my regrets. So I sit by myself and watch dust devils spiral up off the dry ground, driven by the wind, as the light slowly fades.

12

UKRAINE

Igor, as always, is waiting for me at the airport, his head in a tight woollen cap, bobbing up and down in the crowd outside the exit, trying to spot me. His characteristically stern expression breaks into a brief smile when he sees me, but in recent years the smile has become briefer and briefer. His serious enthusiasm, which had so impressed me when we first met, seems to have changed into something grim and rather different.

I awkwardly accept his kisses. We argue over who should carry my suitcase (often full of second-hand surgical equipment) and climb into his van. We drive into the city to the hospital, and Igor holds forth in his broken, staccato English. It was only when I heard Ukrainian spoken by the Ukrainian poet Marjana Savka in Lviv, in the west of the country, that I realized Ukrainian could be very beautiful, and not the rather harsh, declamatory language spoken by Igor.

'Financial crisis terrible. Everybody have money problems. Everybody unhappy. Before crisis my doctors make maybe two thousand dollars in month, now only four, maybe five hundred.' This monologue continues until we reach the hospital. I know that there will be a long queue of patients in the corridor outside the small, cramped office waiting to see me, almost all of them with large and terrible brain tumours and other, often hopeless, neurosurgical problems.

'Two acoustics to see,' he says, as we pass the tall and ugly apartment blocks on the city's outskirts, looking bleak and unwelcoming in the winter mist. There is a thin layer of snow on the ground. I think, not for the first time, how grim Ukraine can be, and how tough its inhabitants have to be to survive. 'Many interesting cases, Henry,' he says happily.

'Well, interesting for you,' I reply grumpily.

'You lose enthusiasm since you retire,' he replies in a disapproving tone of voice.

'Maybe I'm just getting old.'

'No, no, no!' he cries, and then, reverting to his favourite topic, goes on to tell me that twenty Ukrainian banks have collapsed in the preceding year.

We cross the great River Dnieper on one of the many massive bridges built during the Soviet era. The river is frozen, but only in places, and I can see below us the small figures of dozens of people on the shelves of ice, close to the oil-black water, fishing through holes they have cut.

'People drown every day,' Igor observes. 'Twenty this year. It is disaster. It is silly.'

We drive up the steep cobbled street leading from the banks of the River Dnieper to the centre of Kiev and turn onto Institutskaya Street, where a few months earlier dozens of protesters had been killed by snipers in the Maidan demonstration. The SBU hospital where Igor rents space for his private clinic is just round the corner on Lipska Street. The SBU, once known as the KGB, being an important organ of the State, naturally had its hospital in the centre of Kiev. I had been in Kiev on several occasions during Maidan. I spent as much time as I could mingling with the thousands of demonstrators, proud to feel a small part of it.

I had first gone to Ukraine in 1992, just after the collapse of the Soviet Union. Entirely by chance I had met Igor in one of the hospitals I had visited. We became friends and I had been travelling to Ukraine for a few days each year ever since to help him with his surgery. At that time medicine in Ukraine was decades behind the West. I found many second-hand instruments and microscopes for him and taught him everything I knew. At first this was all spinal surgery, and Igor was soon probably the most accomplished spinal surgeon in Ukraine. As his fame spread, more and more patients came to his outpatient clinic with problems in the brain. He badgered me constantly to help him develop brain surgery, insisting that the large and difficult acoustic tumours he had seen me operating on in London could not be treated properly in Ukraine. In Ukraine, because of delays in diagnosis, these tumours are usually very large and the operations correspondingly difficult and dangerous. These are tumours which grow off the hearing nerves within the skull and can become large enough to compress the brain and slowly kill the patient. On my very first visit to Kiev in 1992 I went with two other colleagues from my hospital, one an anaesthetist and the other a pathologist. We visited the major State neurosurgical hospital, one of the two major centres for brain surgery in all of the Soviet Union, where we delivered lectures. My pathology colleague was taken on a tour of the pathology department and came back afterwards to tell us, looking a little shaken, of a series of buckets he had been shown containing the brains of patients who had died after surgery for acoustic tumours.

When Igor and his wife Yelena came to London to Kate's

and my marriage in 2004, all he could talk about was the need to develop acoustic neuroma surgery. Eventually Kate told him that she could stand it no longer. 'Igor,' she said, 'I'm sorry, but we're going to have a moratorium on the word acoustic. We have to talk about something else. We cannot spend every mealtime, every day, with you trying to persuade Henry to show you how to operate on acoustics.' Just for once, Igor did what he was told for a few days. Even I found his intense enthusiasm for neurosurgery and utter commitment to it rather tiring at times. Eventually I agreed to help him, but not without misgivings, as the treatment of patients with brain tumours involves much more than just operating.

In the years before Maidan I would return to England and enthusiastically tell people: 'Ukraine is a really important country!'

This was usually met with a puzzled expression.

'It's part of Russia, isn't it?'

And I would then deliver a little lecture on how Ukraine was one of the great historical watersheds, where Europe met Asia, where democracy met despotism.

I think most of my colleagues and friends in England regarded my slight obsession with Ukraine as an eccentric hobby, but when Maidan started and all of Europe saw the images of the fighting between the demonstrators and the *Berkut* riot police, resembling medieval battle scenes with staves and shields and catapults, and blazing car tyres filling Independence Square with flames and black smoke, I think I could claim some prescience. Igor had had many problems during the twenty-four years we had been working together. He had been something of a medical revolutionary and dissident, using what he had learnt from me to try to improve neurosurgery in Ukraine. The medical system in Ukraine was as authoritarian as the political system and he made many enemies and had many difficulties. But his patients did very well and eventually his clinic became well established. The many attempts by senior colleagues and administrators to thwart him failed. There was something heroic about what he had achieved, and I felt that my work with him over the years was part of the same struggle against corrupt autocracy as the Maidan protests.

There is a turnstile in the small lobby at the entrance to the hospital. The tiled floor is wet with thawing snow brought in on people's boots. Patients and their families can come and go quite freely, but as I am a foreign doctor I am regarded with suspicion by the SBU. When I arrive I have to show my passport to the unsmiling young soldiers behind the glass window next to the turnstile.

'You might be terrorist!' Igor says as the turnstile is unlocked and it clanks in an authoritarian sort of way as I push through it.

'They are SBU soldiers, and hospital has no control of them,' he added.

'It must be an awfully boring job.'

'No, no, no. They are happy not to be at frontline.'

I feel imprisoned once I am inside – imprisoned by my lack of Ukrainian or Russian, and intimidated by the soldiers at the entrance. It is probably quite unnecessary. I once arranged to wait for a film-maker to meet me at the entrance to the hospital. One of the soldiers had to keep me company on the pavement outside until she arrived. When she came she translated what the soldier had started to say to me. I thought he was threatening me with arrest or something similar, but apparently it was a long speech of thanks for my helping Ukrainian patients.

I look back with some shame on the years of my training. It is complete torment to assist a less experienced surgeon than yourself to do a difficult and dangerous operation. Some of the senior surgeons I had worked for simply couldn't do it and left me to get on with it – the so-called 'see one, do one' method of surgical teaching which was one of the more egregious aspects of some English surgical training in the past. I look back with horror on some of the mistakes I made when I was a trainee and, even worse, on some of the mistakes made by my trainees – for which I must hold myself responsible – once I became a senior surgeon myself. But some of my trainers, I now realize, had shown great patience and kindness (and courage) in taking me through operations. I had not thought for a moment how difficult it might be for them, so self-important was I, and so engrossed in what I was doing. Igor, I now realize, is no different. I don't think he ever saw how difficult I found the long outpatient clinics, which could easily last ten or twelve hours, or the agonies I went through as he operated on major brain tumour cases. The more I let him do, the more he would learn, but the greater the risk to the patient and the more anxious I would become. If it seems that it is safe for him to carry on, I retire to the recovery room next to the operating theatre and stretch myself out on the trolley by the window, resting my head on a cardboard box. I would be simultaneously bored and tense, going into the theatre at regular intervals to see how he was getting on and whether I should take over.

'Do you want me to scrub up?' I ask.

'No, no, not yet,' is the usual reply, but sometimes he asks for my help, and sometimes I insist that I take over.

On one previous visit in winter, some years before Maidan, the view through the window as I lay there was uncommonly beautiful – of fine snowflakes drifting down from a grey sky, and of the tall pine and silver birch trees in the hospital courtyard bending with the weight of the snow on their branches. The courtyard itself was virginal white, with only a few footprint tracks on the paths. We were doing a very difficult tumour in a young woman. Both Igor and I thought that I would have to help him, but in the event Igor did almost all of it and she awoke perfectly.

I had passed the long hours by the window watching the snow fall. I thought it was the crowning moment of the many years I had spent teaching him. But on one of my regular visits two years later I heard, only by accident, that the woman had died some months after the operation, from a post-operative infection in her brain. His silence over this, and the way in which he had not sought my advice when she fell ill after the surgery, made me furious and I came close to telling him that I would never return to Ukraine, but eventually I thought better of it. I was told later that he had thought I would refuse to come back to Ukraine if he *did* tell me, which was, of course, the complete opposite of the truth and showed how little he understood me. It reminded me of the long delays before the Soviet government admitted to the catastrophe at Chernobyl.

I told Igor how angry I was, and it was a long time before he grudgingly apologized, but the words of apology seemed almost to choke him, so difficult did he find it to speak them. I went on to tell him how I had once made a similar mistake myself with a post-operative infection, with catastrophic consequences for the patient, and had also failed to ask for help. I still have a photograph of the young Ukrainian woman, which I had taken when I first met her in the cramped little clinic room. She is looking pleadingly at me. Something died within me when I heard of her death, although I still wanted to come out to Ukraine – it had become such an important part of my life. It was only some years later that I understood how wilfully blind I had become. I came to regret bitterly that I had not left him then. I should never have agreed to help him with brain tumour surgery.

13

SORRY

I had gone into work on Sunday as usual. I had been rather depressed about my work in recent weeks and, as I cycled down Tooting High Street in the dark, I decided that it was time to think more positively. Some of my patients came to grief, I told myself, but most did not, and I should remember the successes, not dwell on the failures. I had recently read an article which suggested that stress and anxiety made you more prone to develop Alzheimer's, and also that positive thinking was good for your immune system. I therefore strode into the hospital that Sunday evening full of good intentions.

There were four patients waiting for me – all with brain tumours. I talked to the first three for quite a long time and it was late by the time I reached the fourth. This last patient was a diabetic Asian woman, a few years younger than myself. Her family had brought her to see me two weeks earlier. Her English was limited and the family told me that she had been behaving increasingly strangely over the preceding two years. She had now started to become very drowsy. I had been unable to get much of a history from her, and discussed her problem and its treatment with her family instead. The scan showed that she had a small and benign meningioma at the front of her brain which was producing a great deal of reactionary swelling in the brain, and it was this swelling – medically called oedema – that was causing her symptoms. Surgery would almost certainly cure her. She would return to being the woman she had been, before her brain had started to become oedematous and her personality to change. Brain-swelling can be a major problem with brain surgery for tumours, and it is standard practice to put patients on steroids before surgery to reduce the swelling. In severe cases, such as with this woman, I start the steroids a week before surgery, and I wrote to her GP asking him to do this and warning him that the steroids would make her diabetes worse.

It was ten o'clock at night by the time I came to her room and found her asleep. A little apologetically, I shook her gently. She woke quickly and looked confusedly at me.

'It's Mr Marsh,' I said. 'Is everything OK?'

'Very sleepy,' she said and rolled over away from me.

'Are you OK?' I asked again.

'Yes,' she said, and went back to sleep.

The on-call registrar was standing beside me and I turned to him.

'She was pretty knocked off by the brain-swelling when I saw her two weeks ago,' I said, shrugging. 'And it's late at night, so we'd better leave her alone. The family are in the picture.'

Anxiety is contagious – doctors dislike anxious patients because anxious patients make anxious doctors – but confidence is also contagious, and as I walked out of the hospital I felt buoyed up by the way the first three patients had all expressed such great confidence in me. It allowed me to dismiss the last patient's sleepiness. I felt like the captain of a ship: everything was in order, everything was shipshape and the decks were cleared for action, for the operating list tomorrow. Playing with these happy nautical metaphors, I went home.

The next day got off to a good start. I had slept well and awoke feeling more enthusiastic and less anxious than I usually do on a Monday morning. The morning meeting went well; there were some interesting cases to discuss and I made a few good jokes at the patients' expense, which had the junior doctors laughing. The list got off to a prompt start and the first three tumour operations all went perfectly.

I went into the operating theatre, where my registrar had just positioned the fourth patient on the operating table. As I came in the anaesthetist looked at me with an expression that managed to be both accusing and apologetic at the same time. She had the printout from the blood gas analysis done routinely once the patients are asleep, as doctors say, in her hand.

'Do you know that her blood sugar is forty?'

'Bloody hell!'

'And her potassium is seven. And her pH is seven point two. She must be horribly dehydrated as well. Her diabetes is completely out of control.'

'It must be the steroids – but what was her blood sugar when she came in last night?'

'It seems that the night staff didn't check it. Her blood sugar was only slightly up three days ago when she was seen in the pre-admission clinic.'

'But it should have been checked yesterday anyway, shouldn't it, since she was a known diabetic?'

'Yes. It should have been. I thought she seemed a bit slow this morning when I saw her,' the anaesthetist said, 'but I thought it was because of the tumour, when I realize now it was because she was going into diabetic coma . . .'

'I made the same mistake last night,' I said unhappily. 'But I've never, ever seen anything like this happen before. Presumably we'll have to cancel the case?'

'I'm afraid so.'

'Bloody hell . . .'

'We'll get her round to the ITU and sort out her diabetes – it will take a few days. She needs to be rehydrated. To carry on with the operation now would be hopelessly dangerous.'

'Well, she's had a wacky haircut under GA, even though we haven't removed the tumour,' I said to my registrar as we unpinned her head from the operating table and admired the way he had shaved a couple of inches of hair off the front of her forehead.

'It's very typical of things going wrong in medicine,' the anaesthetist commented from the other end of the operating table. 'It's lots of little things coinciding together . . . If she had spoken better English, if she hadn't already been a bit confused because of the tumour, we'd all have realized that something was going wrong. The failure to check her glucose on admission wouldn't have mattered so much then . . . And if she hadn't been seen in the pre-admission clinic a few days earlier, and instead had just been clerked in and her bloods checked when she was admitted, as we used to do in the past . . .'

'The management introduced the pre-admission clinic as an efficiency measure,' I said.

'But that was because of the lack of beds and increasing workload,' she replied. 'Patients were being admitted later and later on the evening before surgery and there was no time to clerk them in properly.'

'Should we be doing anything about this?'

'I think it's an AIR rather than an SUI,' she said.

'A what?'

'Adverse Incident Report as opposed to a Serious Untoward Incident.'

'What's the difference?'

'An Adverse Incident is anonymous and gets filed somewhere.'

'But where do you send it?'

'Oh, some central office somewhere.'

'But shouldn't I just go and talk to the ward nurses? Do we really want the bloody managers beating them up?'

'Well, you could do that. But it's possible that this is a HONK coma and she might die.'

'What the hell is HONK?'

'Hyperosmolar non-ketotic diabetic coma.'

'Oh,' was all I could say in reply, realizing that my medicine was getting out of date.

So I went to find the nurses. The ward sister was very upset – she is the most conscientious of nurses, and looks perpetually anxious.

'I'll talk to the night staff about it,' she said, looking desperately unhappy. I was worried that she might burst into tears. 'She should have had her glucose checked.'

'Don't get upset,' I said cheerfully. 'These things happen. And the patient hasn't actually come to any harm. Just talk to the night staff about it. Mistakes happen. We're only human. I myself have started an operation on the wrong side you know . . . the important thing is not to make the same mistake twice.'

It had been many years earlier, before we had checklists – an operation on a man's neck for a trapped nerve in his arm. The operation is done through a midline incision, and you dissect down onto one side of the spinal column to drill out the trapped nerve. As I walked down the operating theatre corridor a few hours later, something was nagging me. My heart lurched when I suddenly realized that I had operated on the wrong side. I could quite easily have covered up the mistake – the incision was midline and post-operative scans do not clearly show where you have operated. The pain is not always relieved by surgery, and I could have told the patient a few weeks later that I would have to operate again, without telling him why. I knew of many stories of surgeons lying to patients in similar circumstances. But instead I went to the ward to see the man. It was the old hospital, and he was in one of the few single rooms, which had a window looking out onto the hospital gardens. It was spring, and you could see the many daffodils I had planted some years earlier. I had planted them while in the throes of a passionate – but one-sided – adulterous love affair. It had quickly fizzled out but it laid the foundations for the end of my marriage three years later. I sat down beside his bed.

'I'm afraid I've got some bad news for you.' He looked questioningly at me.

'And what's that, Mr Marsh?'

'I'm terribly sorry but I've gone and operated on the wrong side,' I said.

He looked at me in silence for a while.

'I quite understand,' he then said. 'I put in fitted kitchens for a trade. I once put one in back to front. It's easily done. Just promise me you'll do the right side as soon as possible.'

As I say to my juniors, when you make a stupid mistake, pick your patient carefully.

I went back to the operating theatre and looked up the phone numbers of the various patients' relatives and rang them to report on the day's events. The first three patients were all well: an old man with a pituitary tumour whose eyesight was already better, a monosyllabic garage mechanic on whom I had carried out an awake craniotomy and who had become less monosyllabic now that the operation was over, and a young woman with a tumour at the base of the skull who now had a stiff and painful neck but was otherwise surprisingly well. The last patient was on a ventilator on the ITU having her diabetic coma treated.

So I went home. The day had perhaps been a little chaotic, but none of the patients had come to serious harm. I remained determined that my resolution to see the positive side of neurosurgery should not be broken.

By eleven o'clock I was just about to go to bed when my mobile phone rang. I was upstairs brushing my teeth, and as I had left my phone charging on the kitchen table, I had to stumble naked down the stairs in a hurry, cursing and swearing. Neurosurgeons do not enjoy phone calls in the evening after a day's operating – it usually means that something has gone seriously wrong. The phone had stopped ringing by the time I got to it. The landline phone then started ringing just as the mobile phone also started ringing again with a voicemail message, so I cursed even more as I answered the landline phone.

'The meningioma has blown both her pupils – scan shows severe swelling,' I heard the voice of Vlad, the on-call registrar, telling me.

For a moment I was too surprised to reply.

'But I didn't even operate! I was expecting the other patients to get into trouble . . .'

'Maybe the diabetes, or the rehydration, made the brain swelling worse,' Vlad said. 'What do you want to do?'

'I don't know,' I replied. I sat down on a kitchen chair, stark naked, completely lost for an answer.

'Her biochemistry is OK now. The anaesthetists have corrected it,' Vlad added.

'She's probably had it,' I said after a while. 'But I really don't know what to do . . . if we operate, do a decompressive craniectomy to allow room for the swelling, she might survive but then be left wrecked – and if we don't operate she might just pull through and we can operate at a later date. I don't know what to do,' I repeated.

Vlad didn't comment as I stared vacantly at the kitchen wall.

'It's a unique set of circumstances,' I said. 'It's a toss-up either way.'

I thought of the man with a hammer to whom everything tends to look like a nail.

'We're surgeons,' I went on. 'We tend to see surgical solutions to everything. It doesn't necessarily mean operating is the right thing to do.'

Again Vlad said nothing, waiting for a decision. I sat silently for a while, willing my unconscious to tell me what to do, as the problem was far too unusual for reason and science to give me an answer, although an immediate decision was called for. Vlad was very experienced and shortly to become a consultant himself. The operation was well within his abilities. I could go to bed, I told myself.

'Take her to theatre. Take the front of her skull off and remove the tumour. If the brain is hopelessly swollen leave the bone out.'

'OK,' Vlad said, pleased that a decision had been made and happy at the prospect of operating.

I remained on the kitchen chair for a long time, staring at the kitchen wall. There was no great need for me to go to the hospital, but I realized that I would not get to sleep with the thought that the operation was going on while I lay in bed, so I quickly dressed and drove the short distance, along dark and deserted streets, back to the hospital. I ran up the stairs to the operating theatres on the second floor, but to my annoyance found that the patient had not yet arrived. I went round to the ITU. The patient was lying unconscious on her bed, on a ventilator, surrounded by doctors and nurses.

'Don't bloody wait for the bloody porters!' I said furiously. 'I'll take her to theatre.' So there was the usual hustle and bustle as all the machinery connected to the patient – the syringe drivers, monitors, catheters, IV and arterial lines and ventilator – was disconnected or reconnected to portable equipment and then we set off, a clumsy procession of doctors and nurses, bent double, pulling the bed or pushing or carrying equipment down the long corridor to the operating theatre.

Once there, I had the poor woman's head open within minutes. I rested my gloved finger on her brain.

'Brain's very slack,' I muttered.

'It's probably the ventilation and the drugs she's been given,' Vlad said. 'Look – the brain's pulsatile.' He pointed to the way that the yellow and light-brown mass, covered in blood vessels, that was her brain was gently expanding and contracting in synchrony with the bleeping of the cardiac monitor on the anaesthetic machine. 'She'll be OK.'

'That's what they say,' I said. 'But it doesn't mean much in my experience. When she blew her pupils she might well have suffered catastrophic infarction and most of her brain died. That could be why it's not swollen now. By tomorrow it may well start swelling again as it finally dies off. But she might just pull through . . .' I added, hoping against hope. I picked up a dissector and sucker and it took only a matter of minutes to remove the tumour which had caused all the trouble. It was absurdly easy.

We finished the operation quickly. At the end of these emergency operations, the critical moment is when one anxiously holds open the unconscious patient's eyelids to see if the pupils have started to constrict in reaction to light again. If they constrict, the patient will live.

'I think the left pupil is reacting,' the anaesthetist said happily, peering at the blank black pupil of the patient's left eye. The right eye was hidden by the head bandage I had wound around the patient's head after stitching up the scalp.

I looked. I had to look very closely, my face almost touching the woman's, as I had forgotten to bring my reading spectacles with me in my hurry to get to the hospital.

'I don't think so,' I replied. 'Wishful thinking.'

I wrote a brief operating note, asked Vlad to ring the family once she was out of theatre, and drove home.

I slept badly, waking frequently, hoping against hope, like a rejected lover, that all would be well, that when the dawn came Vlad would ring me to say that she was showing signs of recovery, but the phone remained silent. I went into work next morning and up to the ITU. The ITU consultant was standing beside the patient's bed.

'She's no better,' he said, and launched into a technical account of how he was managing the patient's complicated metabolic problems. He had always struck me as being a bit of a heartless technician.

'I couldn't sleep at all last night,' he suddenly said.

'It's not your fault.'

'I know that,' he replied. 'But it just feels so awful.'

The family were waiting outside the ITU, and we went to talk to them, and prepare them for the worst. There was still some hope, I told them. I said that she might survive but that it was also possible that she might die.

'She might have suffered a catastrophic stroke before I

operated. It's too early to tell,' I told them. I went on to explain that if she had suffered a catastrophic stroke, it would only show up clearly on a scan done the next day. So I said we would get a scan later in the day.

I went in that evening. In the X-ray viewing room I looked at the woman's brain scan on the computer screens. It was mottled and dark – clear evidence of catastrophic damage. Her brain had obviously swollen so badly while her diabetic coma was treated that she had suffered a major stroke. The operation had been too late. I walked round to the ITU where all the family were gathered in the interview room, waiting for me. Their eyes were fixed on me as I told them that there was no longer any hope. I told them that her death had been avoidable, because her blood sugar hadn't been checked on admission. I promised that this would be investigated and that I would report back to them in due course.

As I said this I wanted to scream to high heaven that it was not *my* fault that her blood sugar had not been checked on admission, that none of the junior doctors had checked her over, and that the anaesthetists had not realized this. It was not *my* fault that we were bringing patients into the hospital in such a hurry that they were not being properly assessed. I thought of the army of managers who ran the hospital, and their political masters, who were no less responsible than I was, who would all be sleeping comfortably in their beds tonight, perhaps dreaming of government targets and away-days in country house hotels, and who rarely, if ever, had to talk to patients or their relatives. Why should I have to shoulder the responsibility for the whole damn hospital like this, when I had so little say in how the hospital was run? Why should I have to apologize? Was it my fault that the ship was sinking? But I kept these thoughts to myself, and told them how utterly sorry I was that she was going to die and that I had failed to save her. They listened to me in silence, fighting back their tears.

'Thank you, Doctor,' one of them said to me eventually, but I left the small waiting room feeling all the worse for it.

I left the ITU staff to turn off the ventilator the next day.

I had told the family to sue the hospital – what had happened was indefensible – but they did not. Probably because of my apology.

I wish the authorities responsible for regulating doctors in the UK understood just how difficult it is for a doctor to say sorry. They show little sign of it. The General Medical Council recently produced a document on the Duty of Candour, which is now a statutory obligation. It orders us to tell patients whenever a mistake has been made, both in person and in writing. It would, the document told us, usually be the duty of the senior clinician responsible for the patient to do this, and to apologize, irrespective of who had made the mistake. It went on to add helpfully: 'for an apology to be meaningful it must be genuine,' seemingly unaware of the contradiction between an apology being compulsory and yet at the same time genuine. There was no discussion of how this contradiction can be resolved. It is resolved, of course, if senior doctors like myself feel trusted and respected, and if they have authority, and if they are not compelled to do meaningless things like asking patients to fill in a questionnaire about their behaviour. And if they are given the resources with which to do their work effectively.

I agree with everything in the document about the importance of honesty and apology, but I view with sadness and anger the increasing alienation and demoralization of doctors in England. The government, driven as always by the latest tabloid headlines, has set up an increasingly complex system of bureaucratic regulation based on distrust of the medical profession and its professional organizations. Of course doctors need regulating, but they need to be trusted as well. It is a delicate balance and it is clear to me that in England the government has got it terribly wrong.

14

THE RED SQUIRREL

The two patients with acoustics were waiting for us in Igor's small and cramped office. One was a woman in her fifties, the other in her thirties. Both tumours were very large and both women were starting to lose their balance as a result of the tumours pressing on the brainstem. Patients with tumours this size gradually deteriorate, becoming more and more disabled, and will eventually die – but it can take many years. Surgery, if done competently, has a low risk of killing the patient, but, with the very large tumours, a high risk of leaving them with half their face paralysed, which is very disfiguring and for most people is a life-changing experience.

By this time I had operated on several acoustic tumours with Igor and there had been no disasters, so I was not too troubled about agreeing to operate. The older woman was certain she was happy to proceed with surgery, the younger was very frightened and indecisive. We spent almost two hours talking to her – all of it in Ukrainian, of course, and I spoke little. There was no question that she needed surgery, but she could choose between surgery done by Igor and myself and going to the State Institute. I was in no position to judge how our results compared to theirs.

'What would it cost to have the operation in Germany?' she asked me, Germany being a popular place for wealthy Ukrainians to go for medical treatment.

'At least thirty thousand dollars, probably much more.'

Although she did not say as much, it was clear she could not raise the money to go abroad, but I couldn't deny that probably it would be safer if she did. After two hours of discussion, it was agreed that we would operate. We would do the older patient on Monday and the younger one on Tuesday. There were a few other patients to see, but as it was a Sunday the outpatient clinic was quicker than usual. One was a young woman from a village in the west of the country, with a huge suprasellar meningioma compressing her optic nerves. She had long hair and a very pale face. I was told that she was losing her eyesight, but the details were a little vague.

'She'll go blind without surgery,' I told Igor when I looked at her brain scan. 'But the risks of surgery making her blind are also very high.'

'Oh, I have done several suprasellar meningiomas with good results. You show me how,' he declared confidently.

'Igor, this tumour is enormous. It's the biggest I've ever seen. It's a completely different problem from the usual, smaller ones.'

Igor said nothing in reply, but he looked unconvinced and was obviously itching to operate.

The first acoustic operation next day went well. I cannot remember the details – my memory has been so overlaid by what was to happen later. But I remember that the operation took many hours and, as usually happens when I work with Igor, we were not home until after nine o'clock in the evening.

'It is wonderful when you come!' Igor said to me as we

drove back home, the van bumping over the cobbled street that leads down to the Dnieper. 'It is like holiday for me. I scrub my brain. Recharge my batteries!' I was tempted to reply that my batteries felt correspondingly discharged and flat, and my brain-scrubbing brush worn and bent, but I kept quiet.

I sleep on a sofa bed in the living room of Igor and Yelena's apartment. It's not exactly comfortable but I always sleep well. It is on the sixteenth floor of a typical Soviet apartment block. The sofa bed once opened up in the middle of the night and deposited me on the floor. The view from the window is grimly impressive: a huge circle of identical, shabby high-rise blocks, with a dilapidated school and health centre in the middle of the ring – a Soviet Stonehenge. There is a large flat-screen television in the room, a few religious icons on the wall and a glass-fronted bookcase containing mainly medical books. Like the rest of their apartment, it is very plain and tidy and almost puritanical. Yelena is also a doctor, and works as a cardiologist in the Kiev Emergency Hospital, where I first met Igor in 1992. The family's life is devoted to work.

I usually rise early, woken by the hollow rumbling of the apartment block's battered elevator going up and down as people set off early for work. We get up at six and drive to work forty-five minutes later. The traffic is already very heavy but moves quite quickly, over the high Moskovskyi suspension bridge across the Dnieper. In the west I can see the golden domes of the churches of the Lavra Monastery shining, and to the east the rising sun is reflected in the windows of the garish apartment blocks built in the years of the property boom before the crash of 2007.

We go to see the woman with the acoustic tumour we operated on the day before. She is remarkably well and her face is not paralysed. I am always amazed at how tough the Ukrainians are – she is already standing out of bed, albeit unsteadily. We all laugh and smile happily. In the bed next to her is the young woman with long hair.

'Suprasellar meningioma for surgery tomorrow!' Igor announced. 'The second acoustic patient has sore throat. There is law in this country, if sore throat you not allowed to operate.'

So the day was spent seeing outpatients. In the late afternoon Igor drove me to an empty field on the outskirts of Trojeschina, the bleak suburb of Soviet apartment blocks where he lives. He had bought the field some years ago with a view to building his own hospital there, but then came the financial crash of 2007 and the site remained undeveloped. He was now instead converting an apartment block in the west of the city into a hospital, but had also bought himself a large, unfinished house. I wondered whether his increasingly grim expression was because he had started to overreach himself. The grass was still brown from the winter, burnt black in places, but with a few green shoots to be seen. There was rubbish everywhere. In the distance were the drab buildings of Trojeschina and the tall chimney of a power station. There was a small and dirty stream, partly blocked with plastic bags and tin cans and lined by weary-looking, bare willow trees. Igor produced a breadknife from his jacket pocket and proceeded to cut off a willow branch and stick it in the ground.

'Will that really grow?' I asked sceptically.

'Yes, one hundred per cent,' he replied, waving at the dozen or so willows along the litter-covered bank which had grown over the last ten years.

Opposite us, on the other side of the filthy stream, were a concrete ruin and a restaurant, with piles of rubbish and a dog which barked furiously when it saw me.

'Local people tell me how to do. People cut them down,' he said, pointing to the many stumps and burnt trunks. 'You have to plant five trees if one to live.'

As he planted more willow branches he discussed his endless problems with corrupt bureaucracy.

'This is land of losted possibilities,' he stated. 'First problem is Russia and second is corrupt Ukrainian bureaucrats. Everybody leave,' meaning that young people with any ambitions or dynamism have emigrated. 'I love and I hate this country. It is why I plant trees.'

The next day we operated on the girl who was losing her eyesight. One of the most difficult aspects of working with Igor is that it seems impossible to start the major cases until well after midday. I often complained to him that this was a serious problem as it meant that we ended up doing some of the most difficult and dangerous parts of the operation in the evening, when I, at least, was starting to feel tired. Dangerous and delicate brain surgery can be very intense and intensely draining. But Igor said it was impossible to get the theatre started sooner.

'I must do everything. Check equipment,' he said. 'If nurses or my doctors prepare case, they will make mistakes.'

When I suggested he should delegate more, and that by not trusting his team he was actually creating the problem, he disagreed violently.

'This is Ukraine,' he said. 'Silly people, silly country,' in his characteristically declamatory and assertive style. I had noticed over the years that few doctors or nurses lasted in his department for long. I still don't know if he was right and I was wrong, but I hated these late starts to difficult and dangerous operations.

By one o'clock he was finally starting to open the young

woman's head. I viewed the prospect of the operation continuing until late in the evening with dismay. I braved the guards at the entrance to the hospital and escaped for a walk round a small nearby park, something I have only dared to do since Maidan. The weather had turned suddenly mild and I slithered and slid over melting ice under the dull, grey sky. A red squirrel with high, pointed ears ran in circles around me, occasionally taking fright when I came too close, and darted up the nearest tree. I told the squirrel that I was finding it increasingly difficult to help Igor with these very major operations. I went back to the hospital, showed the guards my passport, and returned to the operating theatre.

The patient's head was almost open and Igor was scrubbed up at the table. I took my usual place on the hard trolley in the corner of the anaesthetic room. The door to the operating theatre was open and I could hear Igor shouting at his staff from time to time. He used a garage compressor for the air-powered drill I gave him nine years ago, used for cutting the bone of the skull to open the patient's head, and the compressor was next to the trolley I was lying on. It went off with a deafening explosion every few minutes when more compressed air was needed. I drifted in and out of sleep, woken at regular intervals by the compressor. It was half past two and they had only just started the bone work! Not for the first time, I told myself that this was the last visit. I had done this often enough before, but I knew it was not a good way to operate . . . And was it really necessary? Did Igor really need to do the occasional difficult case like this? I couldn't go on like this, I told myself, being angry all the time. It was like working in my hospital back in London.

Eventually Igor asked me to take over. He had been slow and careful, but had been unable to find the left optic nerve and needed my help. Once I had settled down in the chair my anger quickly dissipated and the operation seemed to go rather well. I felt happy and concentrated, full of fierce anxiety and excitement – full of the intense joy of operating. It took me several hours of delicate dissection to find the left optic nerve, working in a space only a centimetre wide. When I did, I realized I had wasted my time – it had been so thinned by the tumour that the woman could not possibly have had any vision in her left eye.

'Was she blind in the left eye?' I asked Igor and his staff. All I had been told was that she had only 20 per cent of her vision. But nobody knew. I realized at once that I had made a serious mistake. I should have asked about the nature of her visual loss before the operation. If I had known she was already blind in the left eye I would not have spent hours tiring myself out, trying to find and preserve the left optic nerve

After three hours of intense operating there was a fairly simple lump of tumour left, but I was tired – it was late, and I thought it would be easy for Igor to remove this last part. He had done pretty well so far, although a few of his comments later made me realize that I had not explained the anatomy of the optic nerves as clearly as I should have done. I went off to have a sandwich. When I came back I found that Igor had removed the tumour but had damaged the vital optic chiasm, the area where the two optic nerves meet and cross over. It was entirely my mistake – I should have stayed when Igor was removing that last part of the tumour, or done it myself.

I looked miserably down the microscope.

'She's going to be blind,' I said.

'But right optic nerve OK,' Igor said in surprise.

'But the chiasm has been damaged,' I replied. 'Well, it might well have happened if I had done that last part myself,'

I added. This was certainly true, but at least I would then have felt I had done my best.

Igor said nothing. I don't think he believed me.

We had finished by nine o'clock, and her head had been stitched back together again. She was still anaesthetized when we left but the pupils of both her eyes were large and black and did not react to light – a certain sign of blindness that Igor found hard to accept.

'Maybe she be better tomorrow.'

'I doubt it,' I replied.

Leaving somebody completely blind after surgery – it has happened to me twice – is a peculiarly unpleasant experience. It feels worse than leaving them dead: you cannot escape what you have done. Granted that in all these cases the patients were going to go blind without surgery and had already lost most of their vision, but it is a deeply painful experience to stand next to the patient at the bedside and see their blank, blind eyes fruitlessly casting about. Some of them at first go into a hallucinatory state and think that they can still see, and almost manage to persuade you that they can. You hesitate to disabuse them by demonstrating that they can't – the simplest test is to push your fist suddenly up to their eyes. They do not blink.

We sat at breakfast next morning, as we have done for many years.

'Did you sleep well?' Igor asked me.

'No.'

'Why not?'

'Because I was upset.'

He said nothing.

We made our way to work as usual. I went upstairs with Igor to see the young woman – there was no question that she was completely blind. I found it very hard to look at her while Igor leant over her with a bright desk lamp for a torch, trying to convince himself that her pupils still reacted a little to light, so I left the room.

We returned downstairs to Igor's office to see some of the outpatients already queuing up in the corridor.

'The acoustic woman we cancel is . . . oh!' – Igor waved his arms in the air – 'in terrible way. She hope very much you help with the operation and now you go away.'

After the long conversation and all the negotiations it seemed very cruel that her hopes – exaggerated although they might be – should be dashed and that I would not be involved in her operation after all. Besides, after the previous day's disastrous operating, I felt all the more the need to supervise Igor.

I looked at the diary on my smartphone, resigning myself to the inevitable.

'I can fly back in ten days' time,' I said, 'to do the op. Just for one day, but then I must leave.'

Igor simply nodded his head, and I could not help but feel I was being taken a little for granted. On the way back from work that evening, my guilt and despair about the blind girl finally overcame me. I started shouting angrily at Igor, not blaming him for the operation but for the way I felt he showed no understanding whatsoever for how difficult I found it to help him, and how utterly insensitive he seemed to other people's feelings. I ranted and raved for a while – we were crossing the Moskovskyi Bridge in the dark, and the black waters of the Dnieper, no longer frozen, were below us.

'Oh, I'd better shut up,' I finally said, worried that my outburst would distract him. He had never seen me behave like this before, and I was close to tears. 'Or you'll crash the car.'

'No,' he replied, in his best emphatic and Soviet style. 'I concentrate on driving.' And at that moment I felt an enormous gulf, as wide as the black river below, open up between us. But it was difficult not to be impressed by his apparent calm and detachment.

Later that week I went to the town of Lviv in the west of Ukraine. I had agreed to give a lecture at the Medical School. I spoke of how difficult it is for doctors to be honest. We learn this as soon as we put on our white coats after qualifying. Once we are responsible for patients, even at the lowest level of the medical hierarchy, we must start to dissemble. There is nothing more frightening for a patient than a doctor, especially a young one, who is lacking in confidence. Furthermore, patients want hope, as well as treatment.

So we quickly learn to deceive, to pretend to a greater level of competence and knowledge than we know to be the case, and try to shield our patients a little from the frightening reality they often face. And the best way of deceiving others, of course, is to deceive yourself. You will not then give yourself away with all the subtle signs which we are so good at identifying when people lie to us. So self-deception, I told the Ukrainians, is an important and necessary clinical skill we must all acquire at an early stage in our careers. But as we get older, and become genuinely experienced and competent, it is something we must start to unlearn. Senior doctors, just like senior politicians, can easily become corrupted by the power they hold and by the lack of people around them who will speak truth to power. And yet we continue to make mistakes throughout our careers, and we always learn more from failure than from success. Success teaches us nothing, and easily makes us complacent. But we will only learn from our mistakes if we admit to them – at least to ourselves, if not to our colleagues and patients. And to admit to our mistakes we must fight against the self-deception that was so necessary and important at the beginning of our careers.

When a surgeon advises a patient that they should undergo surgery, he or she is implicitly saying that the risks of surgery are less than those of not having the operation. And yet nothing is certain in medicine and we have to balance one set of probabilities against another, and rarely, if ever, one certainty against another. This involves judgement as much as knowledge. When I talk to a patient about the risks of surgery what I should really be talking about is the risks of surgery *in my hands*, in identical cases, and not just what is stated in the textbooks. Yet most surgeons are singularly poor at remembering their bad results, hate to admit to inexperience and usually underestimate the risks of surgery when talking to their patients. And even if the patient 'does well' and there are no complications after the operation, it can still be a mistake – it may well have been that the patient did not really need the operation in the first place and the surgeon, keen to operate, overestimated the risks of *not* operating. Over-treatment – unnecessary investigations and operations – is a growing problem in modern medicine. It is wrong, even if the patient comes to no obvious harm.

Critical to this is to understand that other people are better at seeing our mistakes than we are. As the psychologists Daniel Kahneman and Amos Tversky have shown, our brains are hardwired to fail to judge probabilities consistently. We are subject to many 'cognitive biases', as psychologists call them, which distort our judgement. We are too biased in our own favour and, under pressure, as doctors often are, we make decisions too quickly. However hard we try to admit to our mistakes, we will often fail. Safe medicine, I told them, is largely about having good colleagues who feel able to criticize and question us. As I said this, I thought of how difficult it is for surgeons like Igor and Dev who work, more or less, on their own.

I was told afterwards that for some of the Ukrainian audience this was almost a life-changing event – to hear a senior doctor admitting to fallibility, and stressing the importance of teamwork, of listening to criticism and of being a good colleague. It was an ironic counterpoint to my increasing problems with Igor.

It was a cold morning and the cars parked in the road outside my house were shrouded in frost that glittered in the moonlight as I bicycled to Wimbledon station nine days later. I sat on the train, wrapped in the heavy overcoat I wear when travelling to Ukraine in the winter, watching the sunrise over the slate roofs of south London alongside the railway line. I lost count many years ago of how many times I have made this journey. In the past I had been excited to return, but now I only felt the intense sadness and regret that you feel at the end of an affair. I felt obliged to keep my promise to operate on the second woman with an acoustic tumour, but I had decided that I could not go on helping Igor with these major cases. He was not the only surgeon doing these difficult operations in Kiev and I was pretty sure that the State Institute – a very large hospital compared to Igor's small, independent clinic – was doing many more such cases, and that it had not stood still since I first visited it twenty-four years earlier. Complex brain surgery, for me at least, is a question of teamwork – having the patient 'on the table' early in the morning, with colleagues and assistants you trust, and with whom you can share some of the burden of post-operative care.

It is a painful truth in medicine that we must expose some patients to risk, for the sake of future patients. As an experienced surgeon I have an ethical duty to the patient in front of me, but also to the future patients of the next generation of surgeons whom it is my duty to train. I cannot train surgeons less experienced than myself without exposing some patients to a degree of risk. If I did all the operating myself, if I instructed my trainees in every move, they would learn nothing, and their future patients would suffer. I had been willing to help Igor do dangerous cases and inflict the torment on myself of supervising his operating, in the belief that he was creating a sustainable and viable future for his clinic and that Ukrainian patients and his own trainees would benefit from it. I also believed him when he gave me to understand that other surgeons in Ukraine could not do these operations. That had probably been true twenty years ago, but I had come to doubt if it was still the case. I had been naive, perhaps worse than that. My own vanity, my wish for what looked like heroic action by working in Ukraine, had distorted my judgement.

I arrived back in Kiev to find that Igor had cancelled the operation on the young woman with an acoustic for a second time. He did not make it entirely clear to me why he had done so. We had a very unsatisfactory meeting with some of his doctors, which I had asked for. I thought I might be able to improve the working relationships in his department by getting them to talk together, but I was wrong. Igor became very angry. He clearly felt that his doctors had no right to criticize him or to complain, and saw the meeting as a plot against him, though of his colleagues' doing, and not mine. And I was a well-meaning but stupid outsider, interfering in a foreign country's internal affairs, having completely failed to understand them.

I returned to London next morning. I subsequently wrote to Igor, trying to explain why I felt unable to go on working with him unless he changed the way he ran his department, so that I felt it had a future, but received no reply. I don't know what happened to the young woman with the acoustic. I had been working with Igor for twenty-four years, for almost as long as my first marriage. In both cases I had clung to the wreckage for far too long, reluctant to open my eyes and admit that my marriage had ended, that my work with Igor no longer had a future. In both cases it was like waking from a nightmare, but one of my own making, and I felt ashamed.

Six months later I returned to Lviv as I had been invited to give some more lectures to the medical students. I talked once again of the importance of honesty and of being a good colleague. But I also told them how essential it is to listen to patients and how difficult it is to learn how to talk to patients as they will rarely, if ever, tell us whether we have spoken well to them or not, for fear of offending us. We get none of the negative feedback and criticism which is such an important part of learning how to do things better. I spoke to them of the importance of telling patients the truth, something most of us doctors find very difficult, as it often means admitting to uncertainty. I told them how the woman with the suprasellar meningioma whom we had left blind had heard I was in Lviv and had asked to come and meet me. I rather dreaded this but when she came, led into the room by her husband, she did not appear especially angry or unhappy. She told me how she had seen many doctors after the operation – it seemed that they had told her that she would have to wait longer for her eyesight to recover, and she wanted to know from me how long this might take.

'What should I tell her?' I asked the students rhetorically. 'I know she is never going to see again. And should she have been informed of that right from the start?'

I told them that it had seemed cruel to deprive her of all hope immediately after the operation, though I had warned her and her husband – but Igor might have chosen not to translate it – that I thought the chances of recovery were very small indeed. But after six months it seemed wrong to continue to lie to her. Up till then in the conversation she had been putting a brave face on things and even making a few jokes about her blindness. But then I told her, slowly, that I wanted her to know how sad I was that the operation had been such a disaster. And now she started crying, and her husband started crying, and I had difficulties not crying myself. And I told her that she never would see again and that she must learn to use a white stick and to read Braille. I delivered a little lecture on neuroscience – about how the visual areas of her brain would quickly be converting to the analysis of sound rather than vision, that blind people could lead almost normal lives, although it was very difficult. And so we talked, and at the end she asked when I was returning to Lviv, as she said she would like to come and talk with me again.

After three months of complete neglect, the weeds in the cottage garden had grown to an extraordinary size – there were stately thistles as tall as young trees, with purple flowers reaching over my head. The cow parsley was ten foot high. There were nameless, numberless plants, some with leaves as large as umbrellas. I was slightly ashamed that I did not know their names. The two rusted corrugated-iron sheds near the lake had almost disappeared under the wild, green tide pushing up against them. The abandoned garden had become an impenetrable jungle. There was a glorious, green freedom to the place, and I felt very reluctant to beat it into submission. But I wanted to know what had happened to the apple trees and single walnut tree I had planted in the winter – they had vanished.

Using my petrol-powered hedgecutter on a five-foot driveshaft, I swept a path towards where I had planted the young walnut tree. At first, to my dismay, all I could see was a dead stalk surrounded by the overbearing, giant weeds – even though I had put black plastic sheeting down around the tree to suppress them. But once I had cleared the surrounding weeds I found, to my joy, that the little walnut tree was alive and well, with large, tender green leaves lower down the stem. I then cut a path to the five apple trees in the opposite corner of the garden, and these I also found to be flourishing – there were even some small apples on their branches.

I spent five hours starting to cut back the weeds and also the overgrown hedge in front of the cottage, which was starting to block the towpath. This was the first hard, physical work I had done for many months and I found, once again, that although exhausting, it was a wonderful panacea. I forgot all my anxieties and preoccupations, I stopped thinking about my future, and my shame, anger and despair over the referendum on Britain's membership of the European Union. The air was full of the green scent of cut grass, the acrid smell of the giant cow parsley and crushed leaves. All I could think about was the next, painful sweep of the hedgecutter, which was balanced on a sling around my neck. My neck clicked and creaked as I worked, and there were constant showers of pins and needles into my right shoulder, from what I assume is a trapped nerve between the third and fourth vertebrae of my cervical spine – the problem has been troubling me for some months. My neck is so stiff that when I try to look up at the stars at night I tend to fall over backwards.

As my body ages, I notice all sorts of new symptoms. My left hip aches a little when I run, my right knee hurts when I sit cramped in airplanes. My prostatism wakes me at night. I am a doctor, so I know what these symptoms mean, and that they will get worse as I get older. I also know that sooner or later I will develop the first signs of a serious illness, which may well be my final illness. I will probably dismiss them at first, hoping that they will go away, but at the back of my mind I will be frightened. I was staying in an expensive hotel recently, and the multiple mirrors in the extravagant marble-clad bathroom not only showed me my elderly, sagging buttocks – a most offensive reminder of my age – but also a mole just in front of my right ear that I had not noticed before. It could not be seen in a single mirror, face-on. I lay in bed, convinced that I had developed melanoma – the most deadly of the skin cancers – and eventually had to get up and search through photographs on my laptop until I found one of myself in profile, which showed that the mole had been present years ago. Only then could I get back to sleep.

I came home from the cottage garden stiff and exhausted, and slept that night for nine hours. I lay in bed in the morning, my neck and back aching, and began to doubt whether I was still capable of all the work required to restore the cottage. I went back later in the day and started to cut out the broken glass from the windows smashed by the vandals. I had spent many hours one year earlier inserting the ogee-shaped glass panes with small nails – known as glazing sprigs – and putty and mastic.

It started to rain very heavily, and the green water of the usually still canal was so flailed by the rain that it seemed to be boiling. The sight distracted me, my hands slipped and I cut my left index finger badly, over the second metacarpal joint, on one of the glass fragments in the window frame, raising a flap of skin over the extensor tendons. It bled profusely, leaving a brilliant red trail on the window frame. I am so used to the sight of blood in the operating theatre that I had forgotten the fairy-tale beauty of its colour. I looked at it in wonder, until the rain started to wash it away. I probably should have taken myself off to hospital to have my finger stitched, but I did not like the thought of queuing for hours, so I went home with a bloody handkerchief wrapped around it. I improvised a repair with a series of plasters cut into strips, and a splint made with outsize matchsticks from an ornamental matchbox a friend had given me.

15

NEITHER THE SUN NOR DEATH

Thirty-five years ago, when I started training as a neurosurgeon, you still had to take what was known as the 'general FRCS'. There was no specialized examination in neurosurgery, and instead you became a Fellow of the Royal College on the basis of an examination that was centred on 'general' surgery, which was mainly abdominal surgery. To qualify for the examination I had to spend a year as a junior registrar in general surgery, which I did in a district hospital in the suburbs of outer London.

It was a busy job, working 'three in seven' – meaning that I was on call in the hospital three nights a week and every third weekend, in addition to working a normal week. You were paid in *umtis* for work done over and above forty hours, an *umti* being a 'unit of medical time', a euphemism whereby four hours' overtime were paid little more than one hour at the basic rate. I was operating most nights – carrying out fairly simple operations for appendicitis or draining abscesses – but usually got enough sleep on which to get by. There were two consultants, both helpful and supportive and good teachers, but – probably like most junior doctors at that time – I took considerable pride in trying not to ask for their help unless absolutely necessary. I therefore learnt quickly, but still look back with deep shame and embarrassment at some of the mistakes I made, when I should have asked for help. At least none of my mistakes, as far I know, were lethal.

I have forgotten most of the patients I looked after during that year, just as I discovered in the Health Camp in Nepal that I have forgotten how to do the operations I did then. One patient, however, I remember very clearly, and even his name. He was a man in his fifties who turned up one evening with his wife in Casualty (as the Accident and Emergency departments were then called). He was smartly dressed in one of those fawn overcoats with a black velvet collar. He and his wife were perfectly polite, but quickly made me aware of the fact that he had previously been a private patient of one of my consultants. He had now run out of insurance and was back on the NHS. They looked very tense, and in retrospect I think they probably had a premonition of what the future might hold for him. He had developed increasing abdominal pain over the preceding two days. I asked him the usual questions about the pain: did it come and go in waves ('colicky' is the medical term), was he still able to pass gas, had he 'opened his bowels' – that clumsy and absurd phrase doctors still use. Had he vomited?

'Yes. That started today,' he said with a grimace, 'and it smelt horrible.' I noted silently that this was almost certainly faeculent vomiting, a sure sign of intestinal obstruction. I asked him to undress and he lay down on the trolley in the curtained cubicle.

His abdomen was criss-crossed with pale surgical scars and distended.

'I had an op for colon cancer three years ago,' he said. 'There were some problems afterwards and I was in hospital for many weeks and needed several more ops.'

'But then he was fine until two days ago,' his wife added, trying to find some grounds for hope. When I palpated – as doctors call examining by touch – his abdomen, I found that it was as tight as a drum. When I percussed it – pressing my left middle finger onto his stomach and then briskly tapping on it with my right middle finger – there was a deeply hollow sound. When I listened to his abdomen with a stethoscope I could hear the 'tinkling, tympanitic' bowel sounds characteristic of intestinal obstruction. He must have waited a long time at home, hoping and praying that the problem would go away.

Something was blocking his gut and the consequences of this are no different from the consequences of a blocked sewer. It is very painful, as the muscular intestines struggle to overcome the blockage.

'We'll need to admit you,' I said, using the reassuring plural that reduces doctors' feelings of personal responsibility and vulnerability. 'We'll get some abdominal X-rays and probably pass a nasogastric tube and put up a drip.'

'Is it serious?' his wife asked.

'Well, hopefully it's just post-op scarring and it will sort itself out,' I replied.

So he was admitted to the surgical ward. The X-rays confirmed intestinal obstruction, showing loops of bowel full of trapped gas. The regime of 'drip and suck' was instituted whereby he was kept nil by mouth and given intravenous fluid, and any fluid in his stomach was aspirated up a nasogastric tube. The idea is to 'rest' the bowel, and sometimes episodes of intestinal obstruction like this can indeed cure themselves without surgery. But this failed to happen and his condition deteriorated. Two days after I had admitted him, my consultant took him to the operating theatre. I was the assistant.

When we opened his abdomen, cutting through the scarred and distended skin and muscle, we found that his intestines were matted together by scar tissue. Parts of the bowel had turned black, meaning that they were dying from strangulation. Untreated, this leads to death within a few days.

'Oh dear. Not good,' my boss said with a sigh. He slowly freed up some of the scarred gut with a pair of scissors so that he could get his hand into the man's abdominal cavity and explore its contents by feeling his way round the liver and kidneys and down into the pelvis.

'Have a feel, Henry,' he said to me. 'Here – up by the liver.'

I put my gloved left hand – I was standing on the patient's left side, my boss on the right – into the gaping, warm hole and felt for the liver. Its normally smooth and firm surface had a large and craggy mass in it.

'A big met,' I said, a met being a metastasis, a secondary cancer which has broken off and spread from the original tumour. The presence of a met usually signifies the beginning of the end.

'Indeed, and there are mets in the mesentery as well. There's not much to do but I suppose we had better resect the gangrenous bits as he might live for a few months yet.'

So we spent the next two hours cutting out the three short, blackened lengths of intestine and then joined – 'anastomosed' – the healthy cut ends together again with stitches.

In brain surgery, if an operation has gone badly you usually know immediately after the operation: the patient wakes up disabled or does not wake up at all. In general surgery, the complications usually occur a few days later, when infections set in or suture lines fail and things fall apart. With this poor man an especially unpleasant complication developed: the anastomoses of his bowel broke down and he developed multiple faecal fistulae through his abdominal wall. In other words, several holes appeared in the operative incision – and in some of the old scars as well – and through these faeces steadily oozed out. The smell was truly awful and there were so many fistulae that it was impossible for the nurses to keep them clean. He had been put in a side room and you had to take a deep breath before going in.

I saw him each day on my regular morning ward round. There was nothing we could do to help him. It was simply a question of waiting for him to die. And he was, of course, wide awake and fully aware of what was happening, slowly dying, surrounded by the terrible smell of his own faeces. He must have seen the involuntary expressions on our faces as we entered his room, steeling ourselves to brave the awful stench.

We started a morphine drip with a pump 'to keep him comfortable', and over a period of many days he slowly died.

I had been a doctor for three years by then but felt utterly unequal to the task of discussing his death with him. I remember him looking sadly into my eyes as I stood above him while the nurses were carrying out the hopeless task of trying to keep his abdomen clean. I don't doubt that we exchanged a few words – probably I asked some banal questions about whether he was in much pain or not – but I know the one thing we did not discuss was his approaching death.

I will never know what he might have said if I had taken the time to sit down beside him and talk to him properly. Was I frightened that he might ask me to 'ease the passing', as it is called in medico-legal language? To increase the morphine and perhaps give him other drugs to end his life? I think it is what I would want if I have the misfortune to end up dying in the way that he died. Or would he have been in a state of denial and still somehow hoping that he might yet live? Or perhaps he would merely have wanted the comfort of talking about his past and memories, or even just about the weather. I remember how my eyes would drift away from him to the window as we talked, to where I could see the autumnal trees at the edge of the hospital car park. I had to make a conscious decision to force them back to look at him.

It is difficult to talk of death to a dying patient, it takes time, and it is difficult if the room stinks of shit. And I know that I let this man down and was a coward.

A few months after qualifying as a doctor, I was working as a medical houseman in a hospital in south London that in the nineteenth century had been a workhouse for the poor. It had not entirely managed to lose the atmosphere of its earlier incarnation. There were long, dark corridors and much of it was very dilapidated, as were so many English hospitals at that time.

It was my duty, as the most junior doctor, to go to the Casualty department and clerk in and admit the patients. If they needed to be admitted to the Intensive Unit, which was immediately above the Casualty department on the next floor, you had to push the patients in their beds a quarter of a mile down the main hospital corridor to the only lift and push them all the way back again along the corridor above to the ITU. If the patient was very ill you tried to run, pushing the bed as fast as you could with a porter and a nurse, but it still took a very long time.

One evening I admitted a man in his fifties with bleeding oesophageal varices. This is a condition when the veins that line the oesophagus – the tube that connects the mouth to the stomach – become enlarged as a result of cirrhosis of the liver. As far as I can remember, his cirrhosis resulted from previous infective hepatitis rather than alcoholism, the commonest cause. The enlarged veins are fragile and can bleed, often torrentially, and the patients come into hospital vomiting large volumes of blood. He was already 'in shock' from loss of blood when I went down to see him. To be in shock is a medical term which means that the blood pressure is falling and the patient will die if the bleeding continues and the blood is not replaced. The patient's pulse will be 'thin and thready', his face pale and his periphery – his hands and feet – cold. He will have an anxious, drawn expression and his breathing will be shallow and fast.

I quickly 'stuck in a large drip'– putting a wide-bored intravenous cannula into a vein in one of his arms – and telephoned my registrar, in some excitement, to say that we had a medical emergency. We transferred him the long distance to the ITU and started to give him blood transfusions. But just as quickly as we poured the blood in, he vomited it all up again. This was almost forty years ago and the only treatment available then, other than blood transfusion, was a device called a Sengstaken tube. This was a very large red rubber tube lined with balloons which you pushed down the patient's throat and then inflated, trying to squeeze the bleeding veins shut. We used to keep them in the ITU fridge, to make the rubber harder – pushing them down the patient's throat and into the oesophagus was very difficult because they were so large, and the patients would cough and gag, simultaneously throwing up blood. We wore aprons and boots. But despite all our efforts he continued to vomit his life away. I remember him silently looking into my eyes as I spent the night with my registrar and the ITU nurses trying to keep him alive. Every few minutes he would feebly try to roll to one side to vomit fresh blood into the grey cardboard vomit bowl beside him.

Despite giving him fresh frozen plasma and clotting factors, we were losing the battle – his blood was becoming visibly thinner and no longer clotting in the vomit bowls. It was quite obvious that he was going to die. My registrar drew up a syringe full of morphine to 'ease the passing'. I cannot remember which one of us administered it. By now it was dawn, and the harsh fluorescent light of the ITU at night was starting to be replaced by the kinder light of the day. Once I had certified him dead shortly afterwards, exhausted, I walked back to my scruffy on-call room, which always smelt of boiled cabbage from the hospital kitchen downstairs. It was summer, and a very beautiful, sunlit morning. That I remember very clearly.

I had been a consultant for only a year. I was responsible for a man with a malignant secondary brain tumour. It was a rare tumour that at first had developed as a solid mass, which I was able to remove. But it was malignant and a few months later recurred. It was now no longer a solid mass and instead individual tumour cells were growing in the spinal fluid, as though the tumour was dissolved in the fluid, making it thick and sticky, causing acute hydrocephalus and terrible headaches. So I operated again, inserting a drainage tube called a shunt to relieve the build-up of pressure in his head. In retrospect this was a mistake, and in later years I would very rarely operate on patients with carcinomatous meningitis, as the condition is called. It is kinder to let the patient die.

The operation was not a great success. His headaches were perhaps better, but he was profoundly confused and agitated. I had informed his family that he was going to die. There was no question of further treatment and I told the family that death was inevitable, which they accepted. But he stubbornly refused to die – his life prolonged, but pointlessly, by the operation. His family became increasingly distressed by the protracted death agony, although the agony was really more theirs than his. One morning the ward sister rang my secretary Gail and asked me to come up to the ward. It was in the old hospital in Wimbledon and my office was in the basement.

'The family are kicking up a terrible fuss,' I was told. 'They are demanding to see you.' So, feeling a little sick with anxiety, and deeply regretting the shunt operation, I walked quickly up the stairs.

It was one of the old Nightingale wards – a large, long room with tall windows and thirty beds arranged in two long rows on either side. The man was in the first bed on the left, and the curtains were drawn. I cautiously put my head between the curtains. The patient's wife was sitting beside the bed, sobbing silently, his middle-aged son standing beside her, his face red with anger. He did not give me a chance to say anything.

'You wouldn't treat your dog like this!' he said. 'You'd put him out of his misery, wouldn't you!'

For a moment I was quite lost for words.

'I don't think he's suffering much,' was all I managed to say, looking at the patient, who was staring silently at the ceiling, mute, awake but seemingly unaware of what was going on around him. I explained that we were giving him heroin (diamorphine is the pharmacological name) through a pump, pointing to the syringe driver bolted to the drip-stand at the bed's head.

'How much longer will this go on for?' asked his wife.

'I don't really know,' I replied. 'Not more than a few days . . .'

'You told us that a few days ago,' his son said.

'All we can do is wait,' I replied.

*

The word 'euthanasia' is used to describe the various ways in which doctors can deliberately bring about the death of patients. It ranges from the criminal mass-murder that went on in mental hospitals in Nazi Germany, to giving painkilling morphine to people in their final agony, to the 'doctor-assisted suicide' that is available in several countries for people faced with certain death from diseases such as terminal cancer or motor neurone disease, or, in a smaller number of countries such as Belgium, for people facing a life of intractable suffering from conditions such as paralysis or incurable depression. In Britain, and in most other countries, doctor-assisted suicide is illegal, although opinion polls in Britain have shown on many occasions a great majority in support of a change in the law. Doctors and members of parliament seem to have more of a problem with it than the public. Doctors rarely admit, even to each other, that they have ever helped patients die – but there can be no question that in the past it happened, and part of me hopes that it still does. You can never know, however, because no doctor wants to talk himself or herself into prison.

A doctor's duty is to relieve suffering as well as to prolong life, although I suspect this truth is often forgotten in modern medicine. Doctors are frequently accused of playing God but, in my experience, the opposite is more often the case. Many doctors shy away from decisions that might reduce suffering but which will hasten a patient's death. It is clear that some people can look death in the face and decide, quite rationally, that their life is no longer worth living. There was, for instance, a young Englishman who had been left quadriplegic by a neck injury while playing rugger, who went to Switzerland to die in the Dignitas clinic. There was an elderly woman who had spent her life looking after psychogeriatric patients and, as she entered old age herself, decided to end her life in the same way rather than run the risk of dementing. I consider these people to have been heroic, and can only hope that I might emulate them should I one day have to face a similar problem. There is no evidence that the moral fabric of these societies that permit euthanasia is being damaged by the availability of this form of euthanasia, or that elderly parents are being bullied into suicide by greedy children. But even if that occasionally happens, might it not be a price worth paying to allow a far greater number of other people a choice in how they die?

One senior politician told me that he was opposed to euthanasia because it would lead to 'targets' – presumably he fears that quotas of elderly people will be encouraged to kill themselves by doctors and nurses. This is an unnecessary anxiety: there are plenty of safeguards in the countries where euthanasia is permitted to prevent this happening. Besides, it is a question of people freely making a choice, *if they have mental capacity*, not of licensing doctors to kill patients who lack capacity. It will not solve the problem of the ever-increasing number of people with dementia in the modern world. They no longer have mental capacity. Besides, doctors do not want to kill patients – indeed most of us recoil from it and all too often go to the other extreme, not allowing our patients to die with dignity.

Clearly the suicidal young – as I once came close to being – need to be helped, as they have all of their lives ahead of them, and suicide is often impulsive, but in old age we no longer have much of a future to which to look forward. It is perfectly rational, on the balance of probabilities, as the lawyers call it, to decide to finish your life quickly and painlessly rather than run the risk of a slow and miserable decline. But neither the sun nor death can be looked at steadily, and I do not know what I will feel as I enter old age, if I start to become dependent. What will I decide if I begin to lose my eyesight, or the use of my hands?

Scientific medicine has achieved wonderful things, but has also presented us with a dilemma which our ancestors never had to face. Most of us in the modern world live into old age, when cancer and dementia become increasingly common. These are now usually diagnosed when we are still relatively well and of sound mind – we can predict what will happen to us, although not the exact timing. The problem is that we are condemned by our evolutionary history to fear death. In the remote past our ancestors – perhaps even the simplest life forms with some kind of brain – did not survive into frail old age, and extra years of healthy life were precious if the species was to survive.

Life, by its very nature, is reluctant to end. It is as though we are hardwired for hope, to always feel that we have a future. The most convincing explanation for the rise of brains in evolution is that brains permit movement. To move, we must predict what lies ahead of us. Our brains are devices – for want of a better word – for predicting the future. They make a model of the world and of our body, and this enables us to navigate the world outside. Perception is expectation. When we see, or feel, or taste or hear, our brains, it is thought, only use the information from our eyes, mouth, skin and ears for comparison with the model it has already made of the world outside when we were young. If, when walking down a staircase, there is one more or one less step than we expect, we are momentarily thrown off balance. The famous sea squirt, beloved of popular neuroscience lectures, in its larval stage is motile and has a primitive nervous system (called a notochord) so it can navigate the sea – at least, its own very small corner of it. In its adult stage it fastens limpet-like to a rock and feeds passively, simply depending on the influx of seawater through its tubes. It then reabsorbs its nervous system – it is no longer needed since the creature no longer needs to move. My wife Kate put this into verse.

I wish I were a sea squirt,
If life became a strain,
I'd veg out on the nearest rock
And reabsorb my brain.

The slow and relentless decline into the vegetative existence that comes with dementia cannot be stopped, although it can sometimes be slowed down. Some cancers in old age can be cured and most can be *treated*, but only a few of us will be long-term survivors, who then live on to die from something else. And if we are already old, the long term is short.

We have to choose between probabilities, not certainties, and that is difficult. How *probable* is it that we will gain how many extra years of life, and what might the *quality* of those years be, if we submit ourselves to the pain and unpleasantness of treatment? And what is the probability that the treatment will cause severe side effects that outweigh any possible benefits? When we are young it is usually easy to decide – but when we are old, and reaching the end of our likely lifespan? We can choose, at least in theory, but our inbuilt optimism and love of life, our fear of death and the difficulty we have in looking at it steadily, make this very difficult. We inevitably hope that we will be one of the lucky ones, one of the long-term survivors, at the good and not the bad tail-end of the statisticians' normal distribution. And yet it has been estimated that in the developed world, 75 per cent of our lifetime medical costs are incurred in the last six months of our lives. This is the price of hope, hope which, by the laws of probability, is so often unrealistic. And thus we often end up inflicting both great suffering on ourselves and unsustainable expense on society.

In every country, health-care costs are spiralling out of control. Unlike our ancestors, who had no choice in these matters, we can – at least in principle – decide when our lives should end. We do not *have* to undergo treatment to postpone fatal diseases in old age. But if we decide to let nature take its course, and refuse treatment for a fatal disease such as cancer, most of us are still faced with the prospect of dying miserably, as in only a few countries is euthanasia – *a good death* – allowed. So, if euthanasia is not permitted, we are faced with the choice of dying miserably now, or postponing it for a few months or longer, to die miserably at a later date. Not surprisingly, most of us choose the latter option and undergo treatment, however unpleasant it might be.

Our fear of death is deeply ingrained. It has been said that our knowledge of our mortality is what distinguishes us from other animals, and is the motive force behind almost all human action and achievement. It is true that elephants can mourn their dead and console each other, but there is no way of knowing whether this means that, in some way, they know that they themselves will die.

Our ancestors feared death, not just because dying in the past without modern medicine must have been so terrible but also for fear of what might come after death.

But I do not believe in an afterlife. I am a neurosurgeon. I know that everything I am, everything I think and feel, consciously or unconsciously, is the electrochemical activity of my billions of brain cells, joined together with a near-infinite number of synapses (or however many of them are left as I get older). When my brain dies, 'I' will die. 'I' am a transient electrochemical dance, made of myriad bits of information; and information, as the physicists tell us, is physical. What those myriad pieces of information, disassembled, will recombine to form after my death, there is no way of knowing. I had once hoped it would be oak leaves and wood. Perhaps now it will be walnut and apple in the cottage garden, if my children choose to scatter my ashes there. So there is no rational reason to fear death. How can you be afraid of nothing? But of course I am still frightened by the prospect. I also greatly resent the fact that *I will never know what happened* – to my family, my friends, to the human race. But my instinctive fear of death now takes the form of fear of dying, of the indignity of being a helpless patient at the mercy of impersonal doctors and nurses, working shifts in a factory-like hospital, who scarcely know me. Or, even worse, of dying incontinent and demented in a nursing home.

My mother was a deeply fastidious person. In the last few days of her life, as she lay dying in her bed in her room in the house at Clapham, with its wood-panelled walls and tall, shuttered sash windows that look out on the Common's trees, she became doubly incontinent. 'The final indignity,' she said, not without some rancour, as my sister and I cleaned her. 'It really is time to go.'

I doubt if she would have wanted to bring her life to a quick end with a suitable pill if she had been given the choice. She strongly disapproved of suicide. But for myself, I see little merit or virtue in the physical indignity which so often accompanies our last few days or weeks of life, however good the hospice care which a minority of us might be lucky enough to receive. Perhaps I am unrealistic and romantic to hope that in future the law in England will change – that I might be able to die in my own bed, with my family beside me, as my mother did, but quickly and peacefully, truly falling asleep, as the tombstone euphemisms put it, rather than incontinent and gasping with the death rattle – at first demonstrating the O-sign, as doctors call it, of the mouth open but with the tongue not visible, to be followed by the Q-sign, which heralds death, with the furred and dried tongue hanging out.

For those who believe in an afterlife, must we suffer as we lie dying, if we are to earn our place in heaven? Must the soul undergo a painful birth if it is to survive the body's death, and then ascend to heaven? Is it yet more magic and bargaining – if we suffer now, we will not suffer in the future? We will not go to hell or linger as unhappy ghosts? Is it cheating, to have a quick and easy death? But I do not believe in an afterlife – my concern is simply to achieve a good death. When the time comes, I want to get it over with. I do not want it to be some prolonged and unpleasant experience, presided over by terminal-care professionals, who derive their own sense of meaning and purpose from my suffering. The only meaning of death is how I live my life now and what I will have to look back upon as I lie dying. If euthanasia is legalized, this question of how we can have a good death, for those of us who want it, with pointless suffering avoided, can be openly discussed, and we can make our own choice, rather than have it imposed upon us. But too often we prefer to avoid these questions, as I did with the poor man at the beginning of my surgical career. It's as though it is better to die miserably than to admit to the inevitability of death and look it in the face.

Once again there is an aneurysm to do. I feel anxious, but also proud that I still have such difficult work to perform, that I might fail, that I am still held to account, that I can be of service. Each time I scrub up, I am frightened. Why am I continuing to inflict this on myself, when I know I can abandon neurosurgery at any time? Part of me wants to run away, but I scrub up nevertheless, pull on my surgical gown and gloves and walk up to the operating table. The registrars are opening the patient's head but I will not be needed yet, so I sit on a stool and lean the back of my head against the wall. I keep my gloved hands in front of my chest with the palms pressed together, as though I were praying – the pose of the surgeon, waiting to operate. Next to me the operating microscope also waits, its long neck folded back on itself, ready to help me. I don't know for how much longer I will feel able to be of use here, or whether I will return, but it seems I am still wanted.

It is hot and dusty outside in the city; the rains are late, the air is yellow with dust and pollution. The haze is so bad that even the nearby foothills are hidden. It gets worse every year. As for the celestial, snow-covered Himalayas, it's almost as though they had never existed. The glaciers are said to be retreating more quickly than in even the gloomiest predictions. It will not be so very long before the rivers run dry.

As I doze, leaning against the theatre wall, I long to return home. I think of the cottage. I walk around the wild, green garden in my imagination. The buds on the young apple trees, a few millimetres in size, are just starting to open, and I can see the miniature petals, tightly furled in layers of pink and white, starting to appear, full of enthusiasm to enter the outside world. It is raining, and the air is wet and smells of spring. The rain forms a million brief circles on the lake, and the two swans are there, looking a little disdainful as always, cruising regally past the reed beds. Perhaps they will make a nest this year among the reeds. I will have made an owl box and put it in the tall willow tree beside the lake. At night

I will practise on the owl whistle I recently bought, hoping to persuade an owl to make its home there. The weeds are starting to reclaim the garden again, but I will have used my book of wild flowers and I will have studied them with great care and learnt all their names. Grass is starting to appear between the red bricks that form the floor of the old pigsty which I spent many days clearing, and weeds are growing back between the cobblestones in front of the drinking troughs. Perhaps the rare vine loved by the foragers will be there as well. I will have brought up the beehives from the garden in London, and I can see the bees coming out of the hives to explore their new home and return with bright-yellow pollen on their legs, now that the winter is past. Perhaps I will buy a small boat, and when my granddaughter Iris is a little older, I will take her rowing on the lake. Even better, perhaps I will make the boat myself, if I have been given the time to build my workshop by then, and all my sharpened tools will be neatly hung and stored, and the place will smell of sawn oak and cedar wood. The windows of the workshop will look out over the lake. In summer there are yellow flags and lilies growing at the side of the lake, just beyond the window. And then there is the derelict, ramshackle cottage, to which I will have given a new life. Perhaps the vandals will have finally left it in peace.

There will be much that needs doing when I return. There will be many things to make or repair, and many things to give or throw away, as I try to establish what I will leave behind. But it seems to me now that it no longer matters if I never finish. I will try not to wait for the end, but I hope to be ready to leave, booted and spurred, when it comes. It is enough that I am well for a little longer, that I have been lucky to be part of a family – past, present and future – that I can still be useful, that there is still work to be done.

ACKNOWLEDGEMENTS

It is only when you write a book yourself that you understand just how important and heartfelt are the acknowledgements. Whatever the quality of this book might be, it would have been many times worse without the comments and encouragement of many friends – in particular Robert McCrum, Erica Wagner, Geoffrey Smith, and my brother Laurence Marsh. My excellent agent Julian Alexander was always on hand to provide wise advice and my wonderful editor Bea Hemming transformed a rather chaotic manuscript into what is, I hope, a proper book. Alan Samson, Jenny Lord and Holly Harley at Weidenfeld & Nicolson gave me further help and advice with the manuscript. I am deeply in debt to my patients and colleagues in London, Kiev and Kathmandu and especially to Upendra and Madhu Devkota, whose exceptional hospitality and kindness make my trips to Nepal so rewarding. I am indebted to Catriona Bass who found the lock-keeper's cottage for me. Most important of all, however, has been the help of my wife Kate, who once again came up with the title and who has been involved in every aspect of the book, as both subject, critic, muse and wife.